BREWING

RSC Paperbacks

RSC Paperbacks are a series of inexpensive texts suitable for teachers and students and give a clear, readable introduction to selected topics in chemistry. They should also appeal to the general chemist. For further information on available titles contact:

Sales and Customer Care, The Royal Society of Chemistry,
Thomas Graham House, The Science Park, Cambridge CB4 0WF, UK
Telephone: +44 (0) 1223 420066
Fax: +44 (0) 1223 423623

New Titles Available

Food – The Chemistry of Its Components (Third Edition)
by T. P. Coultate
Archaeological Chemistry
by A. M. Pollard and C. Heron
The Chemistry of Paper
by J. C. Roberts
Introduction to Glass Science and Technology
by James E. Shelby
Food Flavours: Biology and Chemistry
by Carolyn L. Fisher and Thomas R. Scott
Adhesion Science
by J. Comyn
The Chemistry of Polymers (Second Edition)
by John W. Nicholson
A Working Method Approach for Introductory Physical Chemistry Calculations
by Brian Murphy, Clair Murphy and Brian J. Hathaway
The Chemistry of Explosives
by Jacqueline Akhavan
Basic Principles of Inorganic Chemistry – Making the Connections
by Brian Murphy, Clair Murphy and Brian J. Hathaway
The Chemistry of Fragrances
compiled by David Pybus and Charles Sell
Polymers and the Environment
by Gerald Scott
Brewing
by Ian S. Hornsey

Existing titles may be obtained from the address below. Future titles may be obtained immediately on publication by placing a standing order for RSC Paperbacks. All orders should be addressed to:

The Royal Society of Chemistry, Turpin Distribution Services Limited
Blackhorse Road, Letchworth, Herts SG6 1HN, UK
Telephone: +44 (0) 1462 672555
Fax: +44 (0) 1462 480947

RSC Paperbacks

BREWING

IAN S. HORNSEY

Nethergate Brewery Co Ltd
11–13 High Street, Clare
Suffolk CO10 8NY

ROYAL SOCIETY OF CHEMISTRY

641.23 HORNSEY 1999

Hornsey, Ian S.

Brewing

ISBN 0-85404-568-6

A catalogue record for this book is available from the British Library

Published by The Royal Society of Chemistry, Thomas Graham House, Science Park, Milton Road, Cambridge CB4 0WF, UK

For further information see our web site at www.rsc.org

Typeset by Computape (Pickering) Ltd, Pickering, North Yorkshire, UK
Printed in Great Britain by Athenaeum Press Ltd, Gateshead, Tyne & Wear

Preface

The aim of this book is to provide an insight into, arguably, the most ancient form of biotechnology on this planet. Given the remit regarding the size of the finished article, it has been necessary to be rather selective in a number of subject areas. The responsibility for this selectivity is entirely mine. This project represents the first tome to be written by a practising brewer in these Isles for many years; which, in part, explains the protracted period of time between commissioning and publication of this book. It is somewhat difficult to run one's own business and undertake a creative writing project at the same time. On this point, I must sincerely thank the editorial staff at the RSC for their forbearance.

For my own part, interest in the products emanating from the processes described herein has been extensive, and this fact, together with an innate interest in the technologies themselves, has meant that the task of writing the book has been far from onerous.

I wish to thank my many colleagues in the brewing industry who have made helpful suggestions during the compilation of this work; their contributions have helped greatly. I must stress, however, that any errors in the text are entirely my responsibility.

Finally, I thank my wife for her fortitude and tolerance during the long hours of preparation.

This book is dedicated to my parents.

Ian S. Hornsey

Contents

Chapter 3
Hops

Chapter 7
Microbiology in the Brewing   194

Subject Index   224

Chapter 1

The Definition of Brewing

In the broadest sense the word 'brewing' may be defined as 'The combined processes preparing beverages from the infusion of sound grains that have undergone sprouting, and the subsequent fermentation of the sugary solution produced, by yeast – whereby a proportion of the carbohydrate is converted to ethanol and carbon dioxide.'

The modern connotation of the word would imply 'production of beer', in all its various forms – and this is how the author has interpreted it.

From the definition above it can be inferred that any sound grain (usually graminaceous) can be employed provided that the seed has sufficient polysaccharide food-reserve (endosperm). Cereal grains when raw present a relatively unattractive foodstuff and so a combination of soaking in water, or milling and mixing with water, render products which are far more palatable. These, initially crude, processes have undoubtedly provided the basis for the malting, brewing and baking industries that we know today.

For a variety of reasons barley has become the grain of choice for the brewer, whilst wheat is preferred by the baker.

THE ORIGINS OF BREWING

Enough scientific and archaeological evidence has now been gleaned that it is in order for us to believe strongly that what we now know as 'beer' was first produced in the late fourth millennium BC by the Sumerians in southern Babylonia. The Sumerian civilisation was situated in Lower Mesopotamia – in the alluvial plain between the rivers Tigris and Euphrates – and it was one of the earliest literate civilisations known. The world's oldest recipe, written on Sumerian clay tablets, is for the making of beer; the Sumerians were known to be great beer drinkers. Another early tablet consists of a hymn to the beer goddess, Ninkasi, whose very name means 'the lady who fills the mouth'. According to Cahill,[1] 'Ninkasi was brewer to the Gods themselves', she who 'bakes with lofty shovel the sprouted barley', who 'mixes the *bappir*-malt with sweet aromatics', who 'pours the fragrant beer in the *lahtan*-vessel, which is like the Tigris and Euphrates joined'.

One school of thought attributes the transformation of Man from nomadic hunter–gatherers, to sedentary, crop-growing peoples, to the accidental discovery of the physiologically interesting beverages that resulted from fermented moist wheat and barley. The theory, championed by Dr Solomon Katz of the University of Pennsylvania, propounds that the 'mood-altering' and nutritional properties of these new beverages provided the motivation for a primitive form of agriculture which would have given the populations indulging in it a less strenuous way of life. Dr Katz goes on to propose that the initial discovery of a stable way to produce alcohol provided the stimulus for people to collect different seeds, to cultivate them and try to improve crop characters.

It soon became obvious that air was detrimental to these fermented brews and, thus, one saw the development of narrow-necked storage vessels common in archaeological sites in Mesopotamia. Such vessels, it is surmised, were designed to keep air out and carbon dioxide in.

One of the most thoroughly investigated sites is at Godin Tepe in the Zagros mountains of what is now Iran. There is evidence that the neighbouring Sumerians exploited this area for some of their essential commodities – and brought their beer-making knowledge with them. There were numerous examples of carbonised six-rowed barley excavated together with fragments of pottery jars with unique criss-cross grooving on the inner surfaces. It is thought that these grooves were designed to retain the sediment from the beer after storage. Chemical analysis of sediment found in the grooves indicated the presence of calcium oxalate, a principal (insoluble) component of 'beer-stone'. As the modern brewer knows well, beer-stone is an inorganic, scale-like, deposit that accumulates in fermentation vessels and beer storage tanks. Oxalic acid is present in trace amounts in malt, and combines during the mashing stage with calcium ions to form the insoluble salt. Ancient jars known to have contained wine, cider and mead do not show any evidence of calcium oxalate deposits. The pattern of grooving on the inner surface of the jar fragments bears great resemblance to the Sumerian signs for beer, called kaš (Figure 1.1).

An extensive collaborative study between archaeologists from the

Figure 1.1 *Kaš – early Sumerian signs for beer*

University of Cambridge, and Scottish and Newcastle (now Scottish Courage) Brewery, under the auspices of the Egypt Exploration Society, has produced an insight into how the ancient Egyptians carried out their fermentation technology 3000 years ago. Studies concentrated on two sites on the River Nile: Armana (some 200 miles south of Cairo) and Deir el-Medina; both sites dating to the period known as the New Kingdom (1550–1070BC). Because these sites are outside of the flood-zone of the Nile, the arid climate has allowed desiccated botanical and other biological remains to persist until the present day.

Beer and bread were the most important dietary items of the ancient Egyptians; as evidenced by the plethora of written records concerning the production and consumption of these products. Beer was used as currency at this time, and everyone, from the Pharaoh downwards, drank it. No meal was complete without it and it played a key role in ritual and religious practice, as the number of brewing-related illustrations on the walls of tombs will testify. It has even been suggested that the pyramids were built on a diet of bread and beer! A wide variety of beer types are documented from this period. What are the differences? It is likely that varying categories of beer served the needs of different classes in the population. Archaeological evidence shows that barley was certainly used for brewing, and so was Emmer wheat, which was the dominant cultivated crop throughout much of the Old World. The two grains were, in some products, used in conjunction which could partly explain the variety of beers produced.

With the aid of scanning electron microscopy, Samuel[2] has demonstrated that some grains were sprouted (malted!) before being crushed and used for brewing; the starch grains from such recovered samples showing the characteristic pitting caused by enzymic attack. Unsprouted grains were also used and these appeared to be cooked in hot water before being fermented. There was also evidence of roasted grains being used, presumably to impart colour and flavour to the product. The abundance of lactobacilli from certain brewing sites indicates that these organisms were involved in fermentation as well as yeasts.

As a result of the direct evidence obtained from the area, Samuel proposed a model for New Kingdom brewing (Figure 1.2).

Some of the earliest fermented products in Egypt were very thick in consistency and were called 'boozah', whilst later, slightly more refined, beverages were known as 'hekt'.

The sites around Armana are believed to be within the boundaries of the lost Sun Temple of Nefertiti (*i.e.* these were Tutankhamun's breweries), and in 1996 the Scottish and Newcastle Breweries (now

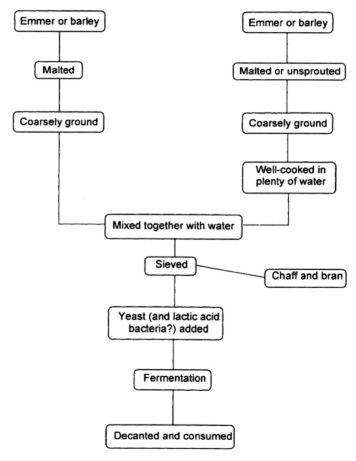

Figure 1.2   *Model for New Kingdom ancient Egyptian brewing*
        (Courtesy of Dr Delwen Samuel)

Scottish Courage), using Samuel's brewing model, re-created this ancient style of beer. Specially grown Emmer wheat was malted for the project, which resulted in a highly distinctive bottled beer called *Tutankhamun Ale*.

Brewing flourished in Egypt until the end of the eighth century AD when Moslem Arabs conquered the region (the Koran forbids the making, sale and drinking of alcoholic beverages), but the art of brewing had spread far beyond the confines of the Middle East; traders to and from the region gleaned the essentials of beer-making and thus the techniques were disseminated. It is to be assumed that it was *via* traders that the beer culture reached the British Isles. Certainly, the Romans found beer to be in production here when Julius Caesar invaded in 55BC. They also found that there were cereal crops under cultivation in certain areas, again emphasising the importance of the sedentary way of life in the gradual civilisation of mankind. At that time, however, mead and cider seemed, according to Roman records, to be far more prevalent beverages over here. The word 'beer' is thought, by some authorities, to be derived from the ancient English word 'beor', which meant 'inferior mead'. Records are scarce from the Dark Ages, but we know that during mediaeval times brewing was more or less confined to monasteries, both in the British Isles and continental Europe. By the thirteenth century there were hundreds of monastic breweries in northern Europe, each supplying the local community with its wares, and this was a convenient way of raising funds for ultimately more saintly purposes! A few of these monastic brewing sites still exist today in Belgium and the Netherlands. Within these religious communities considerable attention was given to improving the quality of the end-product. Many of our present day beer styles originated from these times. For example, Bavarian brewing monks noticed that some yeasts tended to sink during fermentation, and that beers brewed with such yeasts tended to be far more stable than those brewed with yeasts which rose to the top during fermentation. These top-fermenting beers were more prone to infection (by wild yeasts – especially during the summer months when harvesting of crops released vast numbers of airborne microbes), although the nature of the infection was then unknown. The beers brewed with bottom-fermenting yeasts tended to be clearer, slightly fizzy and more stable. Stability could be enhanced if the products were stored at low temperature in vaults or caves, thus enabling brewing to be carried out throughout the year. Such beers were the forerunners of our present day lager styles.

One of the most important results of monastic research was the recognition that hops contained components that were ideal for flavouring and preserving beer. As far as we know, hops were first used by brewing monks in the Hallertau region of Bavaria in 736AD and their use spread gradually throughout northern continental Europe, although they did not reach the British Isles until 1524 *via* Flemish settlers to Kent. The use of hops resulted in a totally different style of beer: these were much more bitter than any preceeding products, and this led to the inception of the 'bitters' that are still prevalent in the UK today. Technically, any product flavoured with the hop is referred to as beer (or bier), whilst products using other flavourings are defined as ales.

LOUIS PASTEUR (1822–1895)

Louis Pasteur (Figure 1.3) was born at Dole in 1822, the son of a tanner. He was appointed Professor of Chemistry at Lille in 1854 and soon became involved in alcoholic and lactic fermentations and the production of vinegar.

His research career spanned the years 1847–1895 and included fields as diverse as crystallography, infectious diseases (anthrax,

Figure 1.3   *Louis Pasteur (1822–1895)*
        (Reproduced courtesy of the Library and Information Centre,
        Royal Society of Chemistry, London, UK)

gangrene and rabies) and diseases of silkworms. He began his studies on fermentation in 1856 when a Lille sugar-beet alcohol manufacturer asked for scientific advice. Studies on the doctrine of spontaneous generation began in 1859 and these led directly to the discovery of the reality of anaerobic life (1861) and the establishment of the fact that fermentations are caused by living organisms.

Research into the causes of wine infections commenced in 1863 and led directly to the laying down of the principles of pasteurisation, whereby bottled wine was rendered more stable by heating to around 55 °C and then allowing it to cool. A similar technique had already been invented by Scheele in 1782 for the sterilisation of vinegar, but Pasteur was actually able to explain the scientific principles behind the process. These extensive researches culminated in the publication of *Etudes sur le Vin* in 1866 and *Etudes sur le Vinaigre* in 1868. Neither work was considered seriously over here because of the inapplicability to British industrial processes and by the lack of any English translation.

The Franco–Prussian war of 1870–1871 and the subsequent defeat of and occupation of parts of France actually provided the stimulus for Pasteur's temporary obsession with beer. Being highly patriotic, he thought that by improving the quality of French beer he could 'get back at the Germans' by producing superior products which would gradually replace the hitherto more consistent and popular German beers from their European markets. French beer production at that time was wholly unscientific and it was pure chance that a wholesome batch would be produced. In 1871, Pasteur visited the laboratory of Prof. Emile Duclaux at Clermont-Ferrand and soon became associated with a small brewery at nearby Chamalieres. There he devised a new method of brewing (which he patented on 28 June 1871). The whole ethos of the new method was to avoid contact between beer and the atmosphere as far as possible, and hence to reduce the likelihood of contamination. According to Pasteur, beers brewed by his new method should be called 'Bières de la Revanche Nationale', or 'Revenge Beers'. The brewery at Chamalieres was very small and Pasteur felt the need to work at a far larger concern. Having no desire to visit a German brewery, he turned to England where he arrived in September 1871 with a small entourage. From his base at the Grosvenor Hotel in Victoria he visited a number of London breweries during his fortnight stay. The only surviving record of his visit is of 9 September 1871 when he toured the Whitbread Brewery in Chiswell Street. Although some major British breweries employed microscopes at this time, Whitbread & Co. did not, and during his investigations (he carried his own microscope) Pasteur found serious contaminations in the porter yeast and the ale yeast. The beer-finings were also found to be contaminated. As a result of a management meeting, fresh yeast was obtained from a nearby brewery and it was agreed that many of Pasteur's brewing tenets should be instigated. On revisiting the Chiswell Street brewery about one week later he found that a microscope had been purchased and new yeast management procedures adopted. Thus, the huge Whitbread tradition of scientific laboratory control of the brewing process emanates from Pasteur's visit.

On his return to France, Pasteur continued his interest in beer, being based at the Ecole Normale in Paris where he established a pilot brewery. Various ammendments and additions were made to his original patent and in 1876 his classic work, *Etudes sur la Bière*, was published. Fortunately this was translated into English three years later by Faulkener and Robb as *Studies on Fermentation* and this led to considerable reaction throughout the British Brewing Industry.

Pasteur's full-time obsession with beer (he did not like it himself) declined by the end of 1876 and he reverted to rather more medical matters. He paid a second visit to London in August 1881 where he represented France at the International Congress of Medicine. (There is no evidence to suggest that he visited any breweries on this occasion.) Although there is evidence to suggest that some of Pasteur's work was not entirely based on original ideas, there is no doubt that he was a major contributor to the science of brewing; indeed, the foundation of Jacobsen's Carlsberg Laboatory in Copenhagen was as a direct result of the Dane's admiration for Pasteur. For the last seven years of his life Pasteur was a member of the Laboratory Club (the forerunner of the Institute of Brewing).

ANTONJ VAN LEEWENHOEK (1632–1723)

Born in Delft, Anton van Leewenhoek was a cloth merchant and trained as a draper. His hobby was lens-grinding, and his inspiration was the English natural philosopher Robert Hooke, whose book *Micrographia*, published in 1665, contained details of small single-lens microscopes. Van Leewenhoek made the best lenses available at the time and his methods were a closely-guarded secret. In 1676 he published his classic drawings of 'animalcules' and communicated a paper on the same subject to The Royal Society in 1677. He was the first person to observe and describe yeasts, when he suspended beer samples in fine glass tubes and observed through his microscope. He thought that the irregular globules he detected (in fact, yeast) were related to and/or derived from starch grains obtained from wheat, oats or barley which were utilised in the manufacture of beer. There was a 'globulist theory' of matter at that time, which stated that 'living things were composed of small globules'. Some of van Leewenhoek's single-lens microscopes still exist (magnification of around × 300).

ROBERT BOYLE (1627–1691)

Boyle, the 'father of modern chemistry' was a pioneer of experimental method and was fully aware of the importance of fermentation. In his 1661 *Essay on the Pathological Part of Physic* he states: '. . . he that thoroughly understands the nature of ferment and fermentation, shall be much better able than he that ignores them, to give a fair account of diverse phenomena of several diseases . . . which will perhaps be never properly understood without an insight into the doctrine of fermentations'. This statement predates by some two hundred years the work of Louis Pasteur (1822–1895).

CHEMISTS AND THE BREWING INDUSTRY

The first trained chemist to be employed in a brewery was Robert Warrington (1807–1867) (Figure 1.4) who joined Truman, Hanbury

Figure 1.4 *Robert Warrington (1807–1867)*
(Reproduced courtesy of the Library and Information Centre, Royal Society of Chemistry, London, UK)

and Buxton of London in 1831. He became an eminent chemist and was elected to The Royal Society in 1864. Only a few breweries had their own laboratories at this time even though the brewing process had become less empirical at the end of the previous century; a fact attributable mainly to the introduction of the thermometer and the saccharometer. The first H.M. Customs and Excise Laboratory was set up in 1842, principally to monitor the increasing trade in the export of beer. Changes in excise laws were made in 1847 when it became necessary for brewers to determine the original gravity of their beers – until then all declarations were made 'on oath'.

By the 1860s, Burton-upon-Trent had become the focus of attention of the brewing industry. This was mainly due to public taste (*i.e.* a swing from the porter-style beers of the London breweries to the pale ale styles of Burton). The growth of the major Burton breweries such as Bass and Allsopps necessitated that production errors be limited, and so we entered the Golden Age in brewing chemistry (see also page 13).

One of the major figures was Cornelius O'Sullivan (1841–1907) (Figure 1.5) regarded by many, as 'The Father of Scientific Brewing'. O'Sullivan studied under Prof. A.W. Von Hoffman at the Royal

Figure 1.5   *Cornelius O'Sullivan (1841–1907)*
(Reproduced courtesy of the Library and Information Centre, Royal Society of Chemistry, London, UK)

School of Mines in London and, upon the appointment of Von Hoffman as Professor of Chemistry in Berlin in 1865, O'Sullivan joined him as his personal assistant. Von Hoffman was then influential in enabling O'Sullivan to obtain the position of assistant brewer and chemist at Bass & Co. in Burton in 1866. He remained at Bass for the rest of his life. O'Sullivan made enormous contributions to the field of carbohydrate chemistry and, amongst other things, he elucidated the nature of maltose (as a disaccharide); he also identified raffinose for the first time and isolated barley gums.

An early associate of O'Sullivan's from his London days was Horace Tabberer Brown (1848–1925) (Figure 1.6), a native of Burton-upon-Trent. In 1866 he became a junior brewer with Worthington's, in Burton, and progressed to the position of manager there in 1873. His early studies revolved around the beer spoilage organism then known as *Saccharobacillus pastorianus* which was a cause of turbidity, but during the course of his career he investigated such areas as brewing water (including the geological aspects); oxygen requirements of yeast; nitrogen in wort and its influence in haze formation; pasteurisation; barley germination and dry-hopping of beers. Brown published numerous papers and made huge contributions to the fields of chemistry, biology and geology.

The last two decades of the nineteenth century saw much scientific brewing research carried out in the British Isles and as a result it became possible to brew all year round. Hitherto, brewing had been

Figure 1.6 *Horace Brown (1848–1925)*
(Reproduced courtesy of the Library and Information Centre, Royal Society of Chemistry, London, UK)

confined to the months of November through to April (attempts to brew in the summer months invariably resulted in severe infection of the fermentation). Beers brewed all year round were referred to as 'running beers', as opposed to 'stock beers' produced during the colder months.

THE CARLSBERG LABORATORY

This unique institution, founded in by J.C. Jacobsen in 1875, was devoted to brewing science – with the proviso that results obtained should be published immediately. No patenting was permitted. Within a few years an enormous amount of fundamental work had emanated from the Copenhagen-based laboratory, notably Johan Kjeldahl's work on nitrogen determination and Emil Hansen's work on pure yeast culture. Hansen (1842–1909) enjoyed an extremely varied career and, apart from his scientific appointments, worked variously as a grocer's boy and house-painter! He was employed by the Ny Carlsberg brewery in 1877 to carry out routine microscopical checks on fermentations. By 1879 he had obtained a doctorate and was appointed Head of the Physiology Department at the Carlsberg Laboratories. Hansen's work led him to conclude that not all brewery infections were caused by bacteria (as Pasteur had surmised), but that yeasts other than *Saccharomyces cerevisiae* were involved. He developed methods for isolating single yeast cells and later sub-culturing them, and this led to his pure culture system being used on a production scale in 1883; this has proved to be the basis for all modern yeast culture protocols.

In the early years of the twentieth century more classic work was carried out by S.P.L. Sørensen on the pH scale; K.J. Linderstrom-Lang on protein chemistry and Ø.V. Winge on yeast genetics. Other research stations had been set up during this period, notably the Faculty of Brewing (Institut für Technologie der Brewerei) at Weihenstephan in Bavaria, which was actually founded in 1865.

There was no equivalent research establishment in the British Isles at that time, although work was being carried out in individual breweries such as Bass, Guinness and Allsopps. The results of such work, however, were not made generally available and the work tended to be aimed at producing solutions to problems encountered within a particular establishment, rather than research *per se*.

Several British academic institutions played a part in the advance of brewing chemistry. The most notable of these were The Royal College of Chemistry (now Imperial College, London); Heriot-Watt College, Edinburgh; Birmingham University and University College, London.

In his excellent article, Anderson,[3] identifies five major eras of brewing research:

(1) The Early Phase (*ca.* 1830–1860), during which period scientists, mostly chemists, became involved with breweries with a view to understanding and solving problems in production.
(2) The Golden Age, which encompassed the late-Victorian period and coincided with booming beer sales.
(3) The Barren Period, which covered the two World Wars and the intervening recession.
(4) The Second Golden Age, commencing in the late 1940s and spurred by post-Second World War optimism and a vast improvement in scientific techniques and general research outlook.
(5) The Post-1970s Period, which, owing to high technology and tighter financial controls, has led to less scope for fundamental research.

SOME IMPORTANT BRITISH BREWING INSTITUTIONS

The Bacterium Club

This was an informal club set up in 1876 by brewers in Burton-upon-Trent. The founders, of whom Horace Brown was the main instigator, were stimulated by the work of Pasteur and, later, Hansen. Cornelius O'Sullivan was one of the founder-members.

The Institute of Brewing

The IOB was originally founded in 1886 as the Laboratory Club, instigated in London by Dr E.R. Moritz, consultant chemist to the Country Brewers Society. It was basically a forum for discussion and dissemination of information, although they published their own Transactions. The Institute was founded in November 1890, after an Extraordinary General Meeting of the Laboratory Club (at which Moritz was not present; he disagreed with the change of name) and soon had three satellite sections: the North (founded 1891); Yorkshire and the North East (founded 1893) and the Midlands (founded 1894). The centres for these sections were Manchester, Leeds and Birmingham, respectively. The four bodies were amalgamated in 1904. There are now seven British sections and three overseas sections.

The Brewers' Guild

In October 1906 a group of Yorkshire brewers formed the Operative Brewers' Guild at a meeting in Leeds. Within five years, membership had increased considerably and sections were formed in all parts of the country. With the publication of a journal, and the concomitant financial responsibility, a limited liability company was established in 1917. For constitutional reasons the name was changed in 1929 to the Incorporated Brewers' Guild and was registered as a company limited by guarantee. The journal became known as *The Brewer*. In January 1993 the name was changed to the Brewers' Guild, and in May 1996 to the International Brewers' Guild. It is now organised into ten British sections and two overseas sections.

Brewing Research International

The formation of the Brewing Industry Research Foundation at Lyttel Hall, Nutfield, Surrey, in 1951 provided the British Isles with its first co-ordinated forum for research and development. It was a joint venture between the Brewers' Society, who provided financial backing, and the Institute of Brewing, with Sir Ian Heilbron as its first director. In 1976 the name was changed to the Brewing Research Foundation and in 1990 it became the Brewing Research Foundation International (BRFI), until October 1997 when its present title was adopted.

The Brewers' Society

The Society was founded in 1904 from an amalgamation of the existing three large brewing interest groups: the London, Burton and Country Brewers' Associations. In 1994 the name was changed to the Brewers' and Licensed Retailers Association.

REFERENCES

1 T. Cahill, *The Gifts of The Jews*, Bantam, Doubleday & Dell, New York, 1998.
2 D. Samuel, *SGM Quarterly*, 1997, **24**, 3.
3 R.G. Anderson, *J. Inst. Brew.*, 1992, **98**, 85.

Chapter 2

Malting

As we have already seen, the use of barley and wheat for malting and brewing was well established by 3000 BC. Raw grains proved to be gastronomically unattractive, but palatability and digestibility were vastly improved by soaking them in water, allowing the grains to germinate and subsequently drying them. As brewing knowledge was disseminated throughout Europe, other grains such as spelt, rye and oats were also used, especially where cultivation of these crops was prevalent. Examination of early monastic brewing sites shows evidence of separate germination (malting) areas for different grain types.

In 1086, the Domesday Book records that the monks of St. Paul's Cathedral brewed 67 814 gallons of ale from:

—175 quarters of barley;
—175 quarters of wheat; and
—708 quarters of oats.

This represents 355 488 lbs (158.7 imperial tons) of grain to brew 1884 barrels of beer!

Even in early times it became evident that beer brewed from barley was easier to produce and superior in quality to that produced from other raw materials. In an attempt to maintain high standards of purity, the Bavarian dukes Wilhelm IV and Ludwig X instigated the German Purity Law (Reinheitsgebot) in 1516. This prescribed that only barley, hops, yeast and water could be used for brewing beer. Some extant German breweries still conform to the Purity Law.

The aim of malting is to transform and mobilise the food reserves in the grain to substrates convenient for mashing in the brewery. It is essentially a controlled germination process and should not be confused with the natural sprouting of seeds in the field, which necessitates vigourous seedling growth at the expense of food reserves. Malt, as prepared, should enable the brewer to produce, after mashing, a wort containing a balance of amino acids, polypeptides, fermentable sugars and essential metabolites sufficient to support yeast growth.

From the end of the sixteenth century onwards barley generally became the preferred grain for malting. There are a number of reasons for this (Figure 2.1):

(1) The seed is surrounded by a husk; an extra protective layer surrounding the actual seed coat, or testa. The husk consists of coarse, dead cells arranged in a honeycomb-like fashion. In the mash tun the husk fragments provide a good filter bed during wort removal. Thus, there are three outer protective layers to the seed (husk, pericarp and testa) which confer vital protection, especially important during storage.

(2) Each grain consists of up to 90% carbohydrate, some 80–85% of which is in the form of starch granules located in the food reserve, or endosperm.

(3) The copious amount of starch present is solubilised and hydrolysed in one operation because there are large amounts of α- and β-amylases present in malted grains.

(4) The gelatinisation temperature of barley starch is relatively low

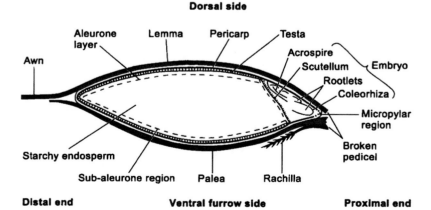

Figure 2.1 *Diagrammatic section through the long axis of a barley corn* (Courtesy of Dr D.E. Briggs, Dr R. Stevens and Dr T.W. Young)

(52–59 °C). This temperature range is below the denaturation temperature for both amylases.

(5) The aleurone layer, which is responsible for producing hydrolytic enzymes, is three cells thick and extremely active. The aleurone layer surrounds the starchy endosperm.

(6) The rootlets and shoot of the embryo, which are extruded from the grain during germination, are very robust and not easily detached. This is especially important during malting when mechanical agitation of sprouting grains is necessary. Wheat embryos are far more readily removed from their seeds.

THE BARLEY PLANT

Botanically, barley falls into the grass family (Graminae) of flowering plants, which in older, natural systems of classification were placed in the sub-class Glumaceae (or Glumiflorae) and, thus, they are related to sedges. Barleys are placed in the genus *Hordeum*, of which there are several species; *H. vulgare* and *H. distichon* being most important in the brewing industry. Within these two main species there are numerous varieties or cultivars. One of the major characteristics of the Glumaceae is that their flowers are enclosed in chaffy scales, called glumes; each individual flower, or floret, being called a spikelet (Figure 2.2). Graminaceous flowers are wind-pollinated (anemophilous) and, therefore, do not possess the conspicuous perianth (the outer parts of the flower) necessary for attracting insects for pollination purposes. The perianth is, in fact, highly inconspicuous and is

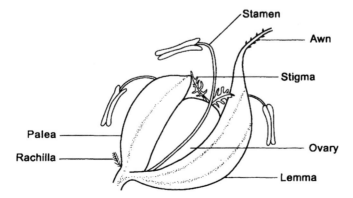

Figure 2.2 *Barley spikelet, with sterile glumes removed*
(Taken from *Brewing Science, Vol. 1*, ed. J.R.A. Pollock. ©
(1979) Academic Press, UK)

represented only by vestiges; the lodicules, which are adpressed to the base of the ovary and only visible after removal of the lemma.

A pair of spike-like, sterile (non-flowering) glumes are situated outside the flowering glumes, which are more membranous and substantial. There are two types of flowering glume. The upper is called the palea, which is situated nearest to the rachis (on the ventral side). The palea has a distinct centrally-situated furrow, which will become the ventral furrow of the barley seed. The lower flowering glume is called the lemma, or inferior palea, and is dorsally situated and partially wrapped around the palea – such that the reproductive apparatus proper is enclosed. The lemma at its distal end (*i.e.* furthest from the stalk) is elogated into the awn. The palea and the lemma will eventually form the husk of the barley seed after fertilisation has taken place. The husk forms some 10% of the weight of the corn and consists mainly of cellulose, although there is some phenolic content which helps to preclude spoilage. Compositionally, the husk remains almost unchanged during malting.

The spatial relationship of lemma and palea gives the seed a discernibly distinct ventral and dorsal aspect (Figure 2.3). (A longitudinal section through the long axis of a barley grain showing the relative positions of the palea and lemma has already been shown on page 16.)

The lemma contains slightly raised lines, called nerves, which indicate the presence of underlying vascular tissue. There is a central median nerve and an inner and outer lateral nerve on either side of the lemma. Nerves may, or may not, be furnished with small teeth called spicules.

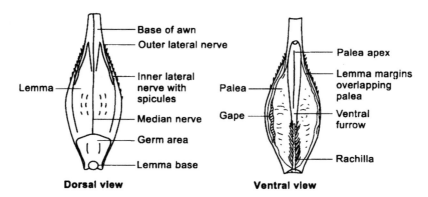

Dorsal view   **Ventral view**

Figure 2.3   *Dorsal and ventral views of a barley seed*
(Courtesy of the National Institute of Agricultural Botany, Cambridge, UK)

Spikelets are arranged in terminal inflorescences, or spikes, commonly called the ear. In reality, the spikes are terminations of the stem of the plant; a major difference being that the distances between nodes are truncated and that at each node three spikelets are produced in an alternate fashion along the axis, or rachis, as it is called in the inflorescence. Each spikelet is attached to the rachis *via* a short flower stalk, or pedicel, which is extended into a small protruding basal bristle called the rachilla.

If all three spikelets at a node (or knot) are fertilised then three seeds will be produced and a six-rowed head will result (*i.e.* six grains at two adjoining nodes) and these are classified as *H. vulgare*. Such barleys are favoured by brewers in the USA and Canada and tend to have a higher enzymic potential than their two-rowed counterparts. If only one of the three spikelets is fertilised then two grains will develop at adjacent nodes and a two-rowed barley ensues. The corns in the latter variety are plumper and more uniform and are preferred by British ale brewers. These plants are placed in the species *H. distichon* and, in effect, what has happened is that only one-third of the inflorescences have developed.

The ovary (Figure 2.4) contains a single ovule (egg cell) which is contained within a covering or integuments; the latter having a small basal aperture, the micropyle, through which the pollen tube can enter and fertilise the ovule. It is at the micropylar end of the grain that the

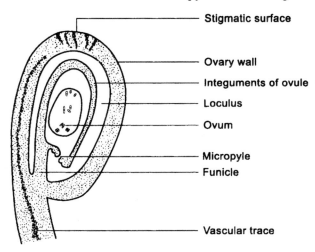

Figure 2.4 *Longitudinal section of the ovary of barley*
(Taken from *Brewing Science, Vol. 1*, ed. J.R.A. Pollock. © (1979) Academic Press, UK

embryo is situated and embryo development takes place. The ovule and integuments are situated in a cavity, the loculus, which is bounded by the ovary wall proper. The ovule is attached to the placenta area of the ovary by a short stalk; the funicle. The ovary wall is subtended at its distal end by a pair of feathery stigmas, which will receive the pollen; predominantly pollen from the same floret.

After fertilisation, the integuments will form the testa of the seed and the ovary wall will be represented as the pericarp.

All barleys are graded by the National Institute of Agricultural Botany (NIAB), in Cambridge, on a scale from 1 to 9; grade 9 being the best, in terms of malting quality. Grade 1 varieties would normally only be fit for animal feed.

Some varieties are planted in September/October and are referred to as 'winter barleys', whilst others are planted in the spring ('spring barleys'). The former are consequently harvested earlier than the latter. All varieties have different agronomical characteristics, disease susceptibilities and yield potential. Winter varieties normally produce greater yields than spring varieties and this helps the farmer to offset the additional costs of winter sowings.

The situation regarding the growing of varieties is ever-changing, with some new ones always being introduced – normally at the expense of the demise of others. Ridealgh[1] concisely summarises the UK malting barley crop for 1998 and pinpoints one or two trends. Of the winter varieties, Maris Otter, that doyen of pale ale malting barleys (and the variety that is used at Nethergate) which was in some danger of disappearing now seems to have stabilised its position, thanks largely to the concerted efforts of certain growers and brewers who would not let the variety die out. Expensive to grow and low yielding varieties, such as Halcyon and Pipkin appear to be on the wane according to Ridealgh, the latter perhaps by the year 2000. Both of these cultivars have been widely grown in recent years. New winters, such as Regina, Fanfare and Gleam seem set to take their place. As can be seen from Table 2.1, these three new varieties comprised almost 35% of the 1998 UK winter barley acreage, with Regina being the most extensively grown winter cultivar – even though it has a tendency to yield high nitrogen levels (see also pages 25–26 and 46–47).

Spring grown crops were again dominated by Chariot and the increasingly popular Optic, the latter being seemingly set to become the predominate spring variety. Table 2.1 also indicates the percentage of each variety that has attained malting grade and the projected demand by the brewing industry.

Table 2.1 *UK malting barley availability – 1998 harvest estimate – pre-harvest*†

Variety	Likely production (tonnes)	% Suitable for malting	Tonnes available for malting	Demand (tonnes)
Halcyon	300 000	60	180 000	220 000
Pipkin	110 000	50	55 000	100 000
Maris Otter	30 000	75	22 500	22 500
Puffin	50 000	30	15 000	55 000
Fanfare	450 000	45	202 500	170 000
Gleam	400 000	50	200 000	100 000
Regina	800 000	40	320 000	100 000
Spice	15 000	75	11 250	11 250
Melanie	250 000	20	50 000	50 000
Other winter	2 395 000	2	47 900	25 000
Total winter	**4 800 000**		**1 104 150**	**853 700**
Chariot	750 000	50	375 000	450 000
Optic	550 000	50	275 000	275 000
Alexis	60 000	75	45 000	50 000
Prisma	200 000	45	90 000	150 000
Cooper	90 000	50	45 000	45 000
Other spring	350 000	30	105 000	120 000
Total spring	**2 000 000**		**935 000**	**1 090 000**
Grand total	6 800 000		2 039 150	1 943 700

† Courtesy of the International Brewers' Guild.

The NIAB produces a list of recommended malting varieties and these then have to be approved by the Institute of Brewing who will only approve those varieties that are available in commercial quantities. Some varieties may only be approved for certain regions in the UK.

World barley production in 1998 was estimated at 139 million tonnes (down from 150 mt in 1997), of which the EEC produced about 52 mt and the UK about 6.5 mt.

The main characteristics of a good malting barley are:

(1) the variety is an accepted one;
(2) the grains are plump with a finely wrinkled surface; no split grains;
(3) there is an even grain size with no foreign seeds;
(4) a good 1000 corn weight;
(5) no pre-germination;
(6) germination very close to 100%;
(7) low nitrogen content (*i.e.* low protein);

(8) no mould or infestations; sound smell;
(9) not dried prior to delivery.

Rapid methods for establishing nitrogen content, moisture and germination capacity may now take only 15–20 minutes, thus enabling assessment of quality to be made whilst the load is being delivered.

THE MALTING PROCESS

Acceptable grains of an appropriate malting quality are taken into the maltings. Their moisture content should be no higher than 20% (15% is preferable). If prolonged storage in the maltings is required then they should be further dried to 12%.

To commence malting the grains are steeped in water to increase their moisture content to a level at which they are capable of germination (normally 42–46%). Traditional (floor) maltings have steeps quite separate from germination areas, but in more modern plants the processes are normally combined. Too little water imparted to the grains at this stage produces weak embryo development (*i.e.* spindly rootlets) and poor modification (and ultimately poor extract in the brewery). Too excessive hydration produces over-modification and high malting losses – even death of the embryo. (See also page 27.)

The grains are not merely left to soak in static water; they are subjected to intermittent periods of drainage (called air breaks) when air is blown through the steep. This enables carbon dioxide to be dispersed, thus promoting germination. Steeping cycles depend upon the sophistication of the equipment available, but a typical cycle would be as follows:

— 8 hour soak (moisture up to 32–34%);
—14 hour drain;
— 8 hour soak (moisture up to 38–42%);
—10 hour drain;
— 8 hour soak (moisture up to 44–46%).

The grains showing the first signs of germination in the form of a slight 'chit' are then transferred to the germination area.

Traditionally, germination would be carried out on floors, with a bed grain depth of about 20 cm. As germination proceeds metabolic heat is produced and this is dissipated by regular (twice per day) raking of the beds with shovels. Raking also prevents matting of the rootlets and carbon dioxide build-up. Ambient temperature should not rise above 19 °C during germination. In floor maltings, which are still used, sufficient cooling can be effected by convection, and sufficient oxygen can be imparted by diffusion because the depth of the grain bed is shallow. If temperature does rise the bed can be spread more thinly; conversely, the bed can be thickened if ambient temperature drops.

First attempts to mechanise malting were made late last century by Galland who drew air down through germinating barley held in a box. His assistant, Saladin, modified the original structure of the box when it was realised that the grains had to be turned as well. The resultant apparatus was named the Saladin Box, examples of which still exist.

In more modern maltings air is blown through the germinating grain bed. Such plants are called pneumatic maltings and, instead of raking, the grains are turned in slowly rotating drums, or by automatic turning arms. Such plants are invariably temperature- and humidity-controlled and grain beds in pneumatic maltings can be anything up to 1.5 m deep. Humidity control is important in order to prevent excessive drying of grains.

Before the late 1940s when refrigeration facilities became available, malting was only possible in the colder months.

The first sign of barley seed growth during steeping is swelling of the grain, and allowance has to be made for this when loading the steep. The first sign of germination proper is the protrusion of the coleorhiza, or root sheath. This is the 'chit' of the maltster and is found at the base of the corn. This will later produce rootlets, or culms. By this time the coleoptile, containing the first leaf, will have penetrated the testa on the dorsal side of the seed. This grows towards the apex of the grain between testa and pericarp. This growth is referred to as the acrospire and is used by the maltster to assess germination rate. When the acrospire is approximately three-quarters of the way along the dorsal side of the grain then the germination process is deemed to be complete and further growth must be arrested; this is effected by kilning, which will be discussed later.

BIOCHEMICAL STRUCTURE OF BARLEY

Whilst germination is manifesting itself to the maltster, a number of very important biochemical changes are occurring within the grain.

These are of ultimate interest to the brewer and are generally referred to as modification processes. Overall, the biochemical structure of a barley seed is very complicated and we shall only concern ourselves with the major constituents, in brewing terms, and the events relating to them.

Starchy Polysaccharides

Starch is located in discrete granules which are found in the endosperm of the seed. There are two sizes: large (20–25 μm diameter) and small (1–5 μm diameter). The latter outnumber the former, but most of the ultimate extract for the brewer will come from large starch granules. All granules are embedded in a protein matrix.

There are two main chemical fractions of barley starch:

(1) Amylopectin, a branched polymer of D-glucose units, linked by $\alpha(1\rightarrow4)$ and $\alpha(1\rightarrow6)$ links. The straight-chain fragments of amylopectin are formed from $\alpha(1\rightarrow4)$ links [Figure 2.5(a)]. Amylopectin comprises approximately 75–80% of barley starch.

(2) Amylose, a straight-chain polymer consisting of $\alpha(1\rightarrow4)$ linked D-glucose units only [Figure 2.5(b)] and comprising some 20–25% of the starch component.

Non-starchy Polysaccharides

These are the sugars, gums and hemicelluloses.

(1) *Sugars*. The main simple sugars in the barley seed are sucrose and raffinose, which are principally located in the aleurone layer and the embryo. The malting process produces far more sugars, both in terms of variety and volume.

(2) *Gums*. These are the β-glucans and pentosans which are soluble in hot water.

(3) *Hemicelluloses*. This fraction refers to the β-glucan and pentosan component that is insoluble in hot water. β-Glucan is a linear polymer of $\beta(1\rightarrow4)$ (70%) and $\beta(1\rightarrow3)$ (30%) linked glucose units. Most of it is to be found in the endosperm cell walls, although a small amount emanates from the husk. Pentosans are, technically, arabino-xylans since they consist of a backbone chain of $\beta(1\rightarrow4)$ linked xylose units which bear side-chains of $\beta(1\rightarrow3)$ linked arabinose units.

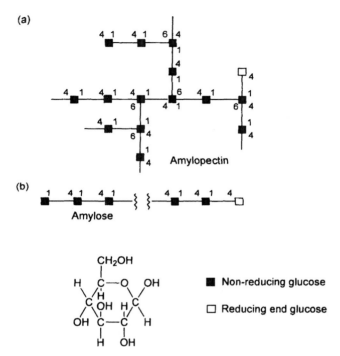

Figure 2.5 *Diagrammatic representation of* (a) *amylopectin, and* (b) *amylose*

Proteins (Nitrogen)

The nitrogen content of a malting barley, which should be in the range 1.4–1.8% dry weight, is a measure of the protein contained therein. In practical terms:

$$\text{Protein content of the grain} = \%\text{N} \times 6.25 \qquad (2.1)$$

The majority of barley N is located in the endosperm as storage protein and enzyme protein. There are four main protein fractions:

(1) albumin (soluble in hot water) and comprising about 4% total protein;
(2) globulin (soluble in dilute sodium chloride); about 31% total protein;
(3) hordein (soluble in 70% ethanol); about 36% total protein;
(4) glutelin (soluble in dilute sodium hydroxide); about 29% total protein.

Albumin and globulin are predominantly enzyme proteins, representing potential sources of β-amylase and peptidases.

Hordein and glutelin are mainly structural proteins, being principally located in starch grain sheaths. They are the main proteins broken down during malting.

In addition to the protein complement, there are a number other nitrogen-containing compounds that are found in small amounts. These include nucleic acids, amines, amides and unbound amino acids.

Lipids

These are located mainly in the embryo and aleurone layer and comprise about 3–4% of the total dry weight of the grain. The predominant fatty acids found in barley lipids are palmitic, oleic and linoleic acids. Phospholipids are also found in minute amounts.

Other Constituents

Monophenols (*e.g.* coumarin) and polyphenols (*e.g.* anthocyanidins) are present in small amounts in the husk, pericarp, testa and aleurone layer.

Various mineral ions are located in the aleurone layer: K^+, PO_4^{3-}, Mg^+, Na^+ and Cl^- are the principal ones. Silica is found in the husk.

The embryo and aleurone layer contain a variety of B-group vitamins, such as biotin and inositol. These will be vital during subsequent fermentation processes.

Main Events during Malting

The unmalted barley grain contains considerable quantities of latent β-amylase, in both soluble and insoluble forms. During malting, the enzyme is totally solubilised.

α-Amylase, on the other hand, is produced during malting as a response to gibberellic acid (*i.e.* gibberellin – mediated α-amylase formation). Gibberellic acid, a natural plant hormone, is produced by the embryo during germination and is transported during the malting process to the aleurone layer where it actually stimulates enzyme production. Apart from α-amylase, gibberellic acid also induces endo-β-glucanases, pentosanases, endo-proteases and limit dextranase. By the end of the malting process the various gibberellic acid-induced lytic enzymes have been translocated to the endosperm where they will cause modification of the texture of the starch from an amorphous mass to a more friable substrate.

A small amount (*ca.* 10%) of starch is hydrolysed during malting (most occurs during mashing). The amylose content rises slightly during malting from 22% (barley) to 26% (malt). Amylopectin is preferentially degraded during malting. The mixture of enzymes capable of degrading starch is known as diastase, and barley diastase will contain a different mixture of enzymes from malt diastase.

From a brewer's point of view, the most important practical aspects of malting are the breakdown of the endosperm cell wall β-glucans (by glucanases) and the subsequent exposure of the protein matrices surrounding the starch grains to attack by proteases. Ideally, some 75% of the β-glucan cell walls should be degraded and about 40% of the protein should be solubilised.

Too much starch hydrolysis during malting leads to excessive embryo growth (manifested by shoots and rootlets) which leads to poor extracts in the brewery. The condition whereby a grain exhibits excessive shoot growth is known as 'shot blade'.

By the very nature of the fact that most lytic enzyme production is gibberellic acid-mediated, the pattern of modification of the endo-sperm is such that the area adjacent to the embryonic (scutellar, or proximal) end of the grain is modified first. Modification then proceeds to the opposite (distal) end of the corn. If malting has been incomplete the distal end can remain unmodified; a condition known as 'hard ends'. To overcome this potential problem, some maltings use an abrasion technique whereby grains are passed through an abrading machine which damages, or scarifies, the pericarp and slightly loosens the husk. This leads to a better distribution of gibberellic acid in the aleurone layer and thus a more even distribution of lytic enzymes. This, in turn, gives a more rapid and regular endosperm modification. If the malt specification permits, abraded grains can be subjected to sprays of low concentration (*ca.* 20 mg kg^{-1}) gibberellic acid solution and this can even further enhance modification. Although efficient under some circumstances, this technique can lead to over-modifica-tion which can present problems during kilning.

Kilning

In most traditional maltings the germinated grain, or green malt, is taken to the kiln where it will be subjected to warm air. Moisture content upon transfer to the kiln would be about 45%. Kilning can take between 16 and 60 hours depending on the plant being used and the type of malt being produced. Temperatures have to be very carefully controlled, the aim being to dry the green malt as quickly as possible without destroying the enzymes produced during malting. These enzymes are most vulnerable when the grain is moist and so early kilning (or 'air-on') temperatures are fairly low (50–70 °C). The actual temperature of the grains at this point will be in the region of 25–30 °C because the greater part of the energy of the air temperature will be absorbed by the latent heat of evaporation. During this phase, enzyme activity, especially of proteases and amylases, is enhanced and colours are developed *via* a reaction of amino acids and sugars which produces the so-called melanoidins. When the malt has reached 15–18% moisture, the proteins will have been rendered stable; this is referred to as the 'break-point'.

After about 10 hours, evaporation slows down and the malt temperature starts to rise in response to gradually rising air-on temperatures (up to 80 °C). At about 20 hours the malt temperature will have reached around 60–65 °C and the moisture content fallen to about 5–8%. Air flow is then reduced and the temperature raised to 100 °C (for an ale malt) and this has the effect of curing the malt. During curing, green malt flavours are lost and the true malty flavours are produced. For an ale malt, this normally takes 6–8 hours, after which the moisture content has been reduced to 2.5–3%. Thermolabile enzymes, such as proteases and glucanases are denatured and any remaining proteins are coagulated. Coagulation of proteins is important for ultimately obtaining beer clarity.

Early kilns were direct-fired, a wide variety of fuels being used depending on the intended use of the malt. Coal would have been widely used, but peat and certain hardwoods were also employed. The use of sulfur (as SO_2) in the kiln contributes to a reduction in malt colour, a rise in soluble nitrogen level and a reduction in nitrosamine formation.

Once the malt has been cured, it is cooled and the rootlets (culms) removed. It is then transferred to silos for storage. As long as the moisture content remains below 4% it may be stored for months under the appropriate conditions.

MALTING LOSS

When barley is converted into malt there is a total loss in weight in the region of 7–10%. There are three main reasons for this: (a) respiratory loss (*ca.* 3–5%); (b) root weight loss (*ca.* 3–4%); (c) loss during steeping (*ca.* 1–1.5%). Some maltings spray germinating grains with dilute solutions of potassium bromate (80–100 ppm) to reduce malting losses (especially root weight loss). Because bromate also inhibits protein breakdown in the endosperm, it is usually used in combination with gibberellic acid (which stimulates proteolysis). Very few brewers now permit the use of bromate in their malt specifications.

Historically, barley and malt have been measured in bushels and quarters, the precise values of which vary in different parts of the world. In the UK the figures are as follows:

—barley bushel	56 lbs	(25.40 kg);
—malt bushel	42 lbs	(19.05 kg);
—barley quarter	448 lbs	(203.21 kg);
—malt quarter	336 lbs	(152.41 kg).

Specification of Malt

Worldwide, brewers can set a whole variety of malt specifications to be met by their maltsters and these would normally include certain tolerances so as to be realistic. Our specifications are relatively straightforward and concentrate on moisture content, colour, extractability, total N content and soluble N content. A sample analysis, which conforms to our specification is illustrated in Table 2.2.

Table 2.2 *Nethergate Brewery pale ale malt specification*

Moisture (%)	3.0
Extract, on sample (litre deg/kg)	301
Extract, dry basis (litre deg/kg)	310
Colour (°EBC, IOB method)	5.2
Nitrogen on dry malt: total (%)	1.50
By Dumas method: total soluble (%)	0.61
Soluble nitrogen ratio (%)	40.7

MILLING AND MASHING

Depending upon the size of the brewery, malt may be conveyed to it in bulk or in bag. Bulk malt is normally screened in the brewery; screening being the removal of undersized grains, stones, *etc.* Bagged malt will have been screened before leaving the maltings.

Before decimalisation, malt was measured in quarters (see above) and the mash tun size was graded according to the number of quarters it could hold.

Milling

Malt, *via* whatever means, is conveyed to a hopper, usually situated directly over the mill, whence it is milled. There are two basic ways of milling: dry and wet.

The standard dry mill has four rollers (two pairs) and is ideal for the well-modified malts that British brewers favour. The two pairs of rollers are separated by a screen and revolving beaters. The first pair of rollers normally receive the grains end-on so that the whole grain is crushed, although the husk survives almost intact. The extruded endosperm and embryos are then broken down by the second pair of rollers to produce grits and flour. Each brewery would have its own specification for the proportions of husks, grits and flour. Grits can be further divided into fine and coarse categories.

Continental brewers, especially lager brewers, who favour less well-modified malts, prefer six roll mills (three pairs of rollers). The pairs of rollers are, again, separated by screens and both well- and under-modified grains can be satisfactorily crushed. All dry mills should be equipped with an anti-explosion device.

Hammer mills and disc mills are also used but mainly for experimental purposes, although hammer milling is employed in some modern breweries where mash filters are installed. In these systems the malt is presented as a fine powder which, whilst yielding very high extraction rates, would cause severe run-off problems in more conventional infusion and decoction brewhouses.

In wet milling systems the grains are soaked in water, in a steeping hopper, such that their moisture content reaches roughly 30%. The steeping hopper is situated immediately above the mill. The damp malt is then passed through crushing rollers where the husks are split open and the endosperm is squeezed out as a slurry. The rollers rotate at high speed and the wet slurry produced is immediately mixed with mashing liquor (water) and passed straight to the mash tun, so, in essence, wet milling and mashing are a combined process.

In a slight variation on this process, the grains may be soaked in water at 80 °C for one minute. This serves to soften the husks and is aimed at giving a more even fragmentation of the endosperm, which will still be in a dry state and easily crushed.

The overall aim of wet milling (apart from the decreased likelihood of explosions) is to increase the extract potential and wort drainage characteristics of the malt in the mash. The latter is effected because the husks will be virtually uncomminuted.

Nethergate dry mills its malts, using a Christy Hunt RBM 2000

Figure 2.6   *Malt hopper, mill and auger (Nethergate Brewery)*

two-roll mill which is situated immediately underneath the malt hopper. An enclosed Archimedes screw (auger) carries grist from the bottom of the mill to the mash tun (see Figure 2.6).

Whatever the process, the general aim of milling is to produce particles of a size that will be rapidly attacked by enzymes in the mash tun. The particles should not be too small, especially with an infusion mash tun, because of subsequent wort run-off problems from the tun. If the particles are too large then enzymolysis is impaired and conversion rates will be slow and incomplete.

Mashing

Mashing, by definition, is the process whereby ground malt, or grist, is mixed with brewing water (liquor) such that a fermentable extract is produced that will support yeast growth, with the subsequent production of beer.

Liquor temperature at mixing (or striking) is critical and in the days prior to the advent of the thermometer, estimation of temperature was very much a rule of thumb. It is claimed that the British brewer used water at a temperature at which his face was best reflected! It is known that water in the range of 65–70 °C exhibits its best reflective properties.

Relatively consistent temperatures could also be achieved by mixing grist with water straight from the well, boiling a fraction of the mix and then re-combining the two mashes. This formed the basis of the (originally continental) mode of mashing called decoction.

In its simplest form, mashing involves a single vessel, the mash tun, in which mixing occurs. Prior to mechanisation, large paddles or oars would have been used manually to obtain an even distribution of grist. The mash would then sit (stand!) for a set period of time and the resultant sugary liquid, called wort, would be removed by various means; often by ladling out. This one-step mashing process is called infusion mashing. In 1853 the first mechanised mashing device was invented by Steel.

At Nethergate, mashing-in (Figure 2.7) is facilitated by grist being carried up from the base of the mill, *via* the auger, to a vertically-situated metal sleeve. The latter encloses an upright mash liquor pipe (Figure 2.8) which is multiperforated at its upper end. Both grist input and rate of introduction of liquor can be regulated to give the desired liquor : grist ratio.

There are other mashing systems, including the wet-milling process, although these are less frequently used, especially in traditional breweries.

In the days before the chemistry of water was fully understood, breweries were solely dependent on their local water source for beer production. Water would be extracted from streams, rivers, bore-holes, springs, *etc.* and the natural chemical composition of such waters would determine the final characteristics of the products. The famous brewing centres, such as London, Birmingham, Edinburgh, Burton-upon-Trent and Munich all became famous for certain beer styles which resulted from distinctive brewing waters. Nowadays, more or less any water can be chemically modified to produce liquor suitable for brewing whatever beer style is required. It is in the mash tun that water composition, and especially its pH, plays the most critical role. For a comprehensive account of the significance of water in brewing, the reader is recommended to read Moll.[2]

The ideal ionic composition of a brewing liquor will vary according to the beer style required.

Table 2.3 shows the ideal characteristics of liquor for five beer styles.

The hardness of water is of prime importance to the brewer, especially during mashing. There are two main categories of hardness: permanent and temporary. The former is caused by $CaSO_4$, CaO, $Ca(NO_3)_2$, $MgSO_4$, MgO and $Mg(NO_3)_2$, whilst the bicarbonates of

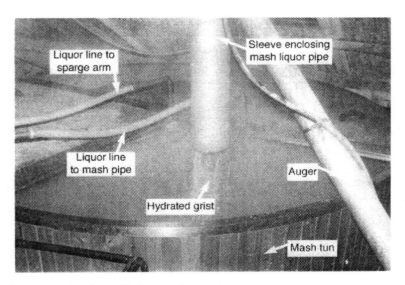

Figure 2.7 *Mashing (Nethergate Brewery)*

Figure 2.8 *Mash apparatus showing liquor pipe (Nethergate Brewery)*

Table 2.3 *Liquor profiles for five different beer styles*[†]

	Pilsen (mg l^{-1})	Burton-upon-Trent (mg l^{-1})	Munich (mg l^{-1})	Dortmund (mg l^{-1})	Vienna (mg l^{-1})
Ca^{2+}	7.1	268.0	75.8	262.3	162.8
Mg^{2+}	2.4	62.1	18.1	22.9	67.6
HCO$_3$$^-$	14.0	280.1	151.5	282.4	242.9
SO$_4$$^{2-}$	4.8	638.3	9.6	289.2	216.3
NO$_3$$^-$	Traces	31.0	Traces	Traces	Traces
Cl$^-$	5.0	36.0	2	107	39.0
Total hardness (°F)	0.9	29.0	8.2	23.5	21.6
Temporary hardness (°F)	0.7	7.4	8.0	9.4	17.3
Permanent hardness (°F)	0.2	21.6	0.3	14.1	4.3
Residual alkalinity (°F)	0.5	0.2	5.9	3.1	4.4

[†] Courtesy of the Institute of Brewing.

calcium and magnesium constitute the latter. Temporary hardness is one of the main potential problems in a brewing liquor, since dissociation of bicarbonates results in an increase in pH in the mash tun. One of the commonest forms of modern liquor treatment encompasses an acid (often sulfuric) to remove temporary hardness, accompanied by gypsum to increase permanent hardness.

Total hardness of a water sample encompasses both categories mentioned above and is normally calculated by titrating 100 ml water with 0.02 M EDTA to indicator end-point. Different countries have defined hardness in a variety of ways and, accordingly, a number of units exist, principally:

—France, where 1 °F ≡ 10 mg l^{-1} CaCO$_3$;
—Germany, where 1 °D ≡ 10 mg l^{-1} CaO;
—UK, where 1 °E ≡ 14.3 mg l^{-1} CaCO$_3$;
—USA, where 1 °USA ≡ 1 mg l^{-1} CaCO$_3$.

Temporary hardness can be partially removed by boiling, whence insoluble CaCO$_3$ is deposited, whilst addition of lime brings about the same result, although time is needed for the precipitate to form. Many brewers now use specially formulated acid liquor treatments to adjust hardness.

Also of interest to the brewer is the hardness due solely to calcium ions which can be calculated by incorporating a specific Ca-binding indicator with the EDTA titration. By definition, magnesium hardness represents the difference between total hardness and calcium hardness.

Hardness values are normally expressed in millivals (millimoles divided by the valency of the element in question) per litre.

Measurement of alkalinity, with its obvious ultimate effect on wort pH, can be carried out by titrating 100 ml water with 0.1 M HCl using the methyl orange (pH 4.4) and phenolphthalein (pH 8.3) end-points. The methyl orange titration accounts for hydroxides, carbonates and bicarbonates of the alkali and alkaline-earth metallic elements, whilst tiration with phenolphthalein takes into account half of the carbonates present and all of the hydroxides. Titration with methyl orange, in essence, measures total alkalinity.

In 1941, Kolbach[3] introduced equilibrated alkalinity and expressed it *via* equation (2.2) as:

$$\text{Equilibrated alkalinity} = \frac{\text{calcium hardness}}{3.5} + \frac{\text{magnesium hardness}}{7}$$

$$(2.2)$$

This enabled residual alkalinity to be calculated as total alkalinity minus equilibrated alkalinity.

In general, the principal ions present in most brewing liquors are HCO_3^-, CO_3^{2-}, Cl^-, SO_4^{2-}, K^+, Na^+, Ca^{2+} and Mg^{2+}. There are also small amounts of trace elements present. Of the above ions, HCO_3^-, CO_3^{2-} and Ca^{2+} may be considered the most important, especially in terms of their ability to influence pH.

The most desirable effect of Ca^{2+} is associated with its ability to react with PO_4^{3-}, present in the malt to form the tertiary compound $Ca_3(PO_4)_2$, which is relatively insoluble. This reaction leads to the release of H^+ which lowers the pH of the mash. The equation is as follows:

$$H_3PO_4 \underset{H^+}{\overset{H^+}{\rightleftharpoons}} H_2PO_4^- \underset{H^+}{\overset{H^+}{\rightleftharpoons}} HPO_4^{2-} \underset{H^+}{\overset{H^+}{\rightleftharpoons}} PO_4^{3-} \qquad (2.3)$$

(primary)	(secondary)	(tertiary)
dihydrogen phosphate	monohydrogen phosphate	phosphate

Phosphates in malt emanate from phosphate esters of inositol and phytins.

Magnesium ions do not have quite the same effect during the mash, principally because $Mg_3(PO_4)_2$ is more soluble than its calcium counterpart.

The most undesirable effects of carbonates and bicarbonates can best be demonstrated by equation (2.4):

$$CO_2 + H_2O \rightleftharpoons H_2CO_3 \underset{H^+}{\overset{H^+}{\rightleftharpoons}} HCO_3^- \underset{H^+}{\overset{H^+}{\rightleftharpoons}} CO_3^{2-} \quad (2.4)$$

In terms of the mash pH, it is necessary for the reactions to proceed from left to right, which would yield H^+ and contribute to a low mash pH. In practical terms, the reactions proceed from right to left which means that H^+ ions are consumed; thus, there is a tendency for the mash pH to rise.

The pH for mashing liquor should be such that the net pH after mashing is 5.4. This is the optimum pH for amylolytic activity and will, therefore, produce maximum levels of starch breakdown (saccharification).

As a result of mashing some 90–95% of the malt starch should be solubilised and converted into fermentable sugars. Malt itself contributes to the mash pH; dark malts produce lower pH mashes than pale ale malts, which, in turn, produce lower pH mashes than lager malts.

The optimum pH of the proteolytic enzymes present in malt is in the range 4.5–5.0 and so they do not work as efficiently as amylolytic enzymes. Consequently, only some 35–40% of malt protein is solubilised during the mashing stage.

A reduced mash pH causes low amylase activity and problems with wort run-off (called 'setting' of the mash), whilst a raising of pH causes extraction of phenolic substances which gives the final beer a harsh (astringent) character and can produce hazes.

The mash has its own intrinsic buffering capacity contributed to by phosphates, amino acids and peptides in the malt, and so, provided that the initial pH of the mashing liquor is as required, a suitable, sustainable pH can be attained throughout the mash period.

For an infusion mash the normal net temperature should be in the range 64–65 °C. In order to achieve that temperature in the mash tun, the temperature of the brewing liquor being mixed with the grist must be precise. This is called the striking temperature and it will vary according to the characteristics of the malt being used. Especially important is the moisture content of the malt. Malt which has re-absorbed moisture after kilning is referred to as being 'slack'. The higher the moisture content, the higher the striking temperature required. This can prove to be deleterious to enzymes in the malt and lead to subsequent poor conversion of starch and proteins (which leads to hazes).

Proteolytic enzymes have lower temperature optima (50–55 °C) than do amylases (60–68 °C). The amylases themselves have slightly different optimum temperature ranges for their activity:

—α-amylase works most efficiently in the range 64–68 °C;
—β-amylase works most efficiently in the range 60–65 °C.

Thus, maximum amylolysis occurs in the range 64–65 °C.

Decoction mash processes involve the use of three separate vessels:

—a mash vessel, where liquor and grist are mixed;
—a decoction vessel, where heating takes place (often called the mash copper);
—a filtration vessel – the lauter tun.

This process is used for brewing with the less well, or irregularly modified malts favoured by continental brewers, or for grists with high levels of adjunct (materials other than malt) – (see page 51).

Mashing-in usually occurs at ambient temperature to enable soluble components from the grist to be extracted. Hot liquor is then added to the mash and mixed to bring the temperature up to 35–40 °C. One-third of the mash is then removed to the mash copper, heated, and held at 65 °C for about 20 minutes (this enables starch conversion). The mixture is then brought to the boil and held at that temperature for between 15 and 45 minutes (depending on style of beer being brewed; longer boiling imparts more colour). This fraction is then returned to the mash-mixer where it will raise the overall temperature to around 50–52 °C. One-third is then removed to the copper, boiled, held for a short period and then returned to the mash. This will raise the mash temperature to around 65 °C, whence amylolysis can occur. After a set stand period, a final fraction is pumped to the copper, heated and returned. This will raise the mash temperature to 76 °C. No further enzyme activity is now possible and the mash is then pumped to the lauter tun for filtration. The lauter tun is equipped with internal revolving blades to assist wort run-off, because most of the air will have been removed as a result of the decoction and pumping processes.

There are many variations, in terms of temperature and time regimes, and the decoction mash process enables a wide range of raw materials to be used and beers to be produced. For more well-modified malt grists with low adjunct levels, a quicker double-decoction system can be employed.

Decoction mashes are usually stirred by means of slowly rotating blades in order to achieve maximum extraction. Stirring causes air to be lost from the mix. Infusion mashes are not generally stirred, so the air contained in the mash remains *in situ*.

Because the mash temperature is gradually being raised by decoction, some of the enzymes such as proteases and β-glucanases, with lower temperature optima than the amylases, have a chance to operate for a while at their optimum temperatures. It should be realised, however, that the boiling of fractions of the mash causes protein (enzyme) inactivation.

The grain bed depth in a lauter tun is shallow (rarely more than 0.5 m) and so, with the blades rotating, wort run-off is rapid.

The grain bed depth in an infusion mash will be in the region of 1–2 m and, in effect, the grains are 'floating on a cushion of trapped air'. The stand period of an infusion mash will vary according to the brewer's requirements and can be anything from 30 minutes to 4 hours. A standard mash tun will be equipped with a series of plates at the base of the vessel, which are actually an inch or so above the bottom of the vessel. These plates, called false-bottom plates (Figure 2.9), are furnished with grooves or other perforations to give porosity. Wort from the mash is drawn through these perforations and subsequently out from the bottom of the mash tun. The rate of wort run-off is dependent upon the depth and consistency of the mash and the degree of perforation of the false-bottom plates. Extensive accounts of infusion and decoction mashing are provided by Briggs and co-workers.[4]

Mash Separation

Regardless of the mode of mashing, mash separation is important for three main reasons:

(1) to get maximum extraction of soluble fermentable sugars;
(2) to obtain bright worts with minimum suspended solids;
(3) to minimise dissolved oxygen concentration (dO_2) in the wort.

Secondary considerations must be:

Figure 2.9 *Mash tun showing false bottom plates in upright (cleaning)
position (Nethergate Brewery)*

(a) to turn around the brewhouse as quickly as possible;
(b) to minimise moisture content of spent grains;
(c) to minimise effluent production.

A wide range of equipment has been used for mash separation
purposes, and includes:

—the mash tun;
—the lauter tun;
—the Strainmaster lauter tun (patented by Anheuser Busch Inc.);
—the mash press;
—the high pressure mash press;
—the mash filter.

Of these, only the mash tun, lauter tun and mash filter are used to any
great extent.

Two key physical processes are involved in mash separation:
leaching and filtration. Leaching is the dissolving of solids from grain
particles and the diffusion of such dissolved solids into the liquid
phase of the mash. Filtration is the separation of liquids and solids
through a filter bed. Leaching is driven by a concentration gradient

and will be maximised by maintaining that gradient, *i.e.* when wort is drawn off it must be replaced by more liquor (sparging, see below). The rate of leaching is governed by particle size; thus, the smaller the grist fragments the less diffusion distance is required (high surface to volume ratio) and therefore leaching is enhanced. Whilst smaller fragments yield better leachate (extract), if present in over-abundance they cause run-off problems – especially in a mash tun because of an increase in resistance to liquid flow.

After withdrawing a specified volume of wort (at this stage called sweet wort) from the mash tun, more liquor is sprayed over the mash *via* a rotating arm (Figure 2.10). The temperature of this liquor will be slightly higher than that used for the initial mash (in the region of 70–75 °C) and the process is called sparging, the rotating arm being called the sparge arm. The aim is to extract and drain as much goodness out of the mash as possible and to replace the liquid being drained from the mash. The elevated temperature of the sparge liquor means that the net temperature in the mash can be maintained and any persisting enzymes are now destroyed.

The first runnings from the mash tun will have a high specific gravity (1070–1080° for ordinary strength beers) and will be very viscous, whilst at the completion of sparging the specific gravity of the final runnings can be in the order of 1005–1010°. In some breweries it is the practice to re-circulate the first runnings back over the mash

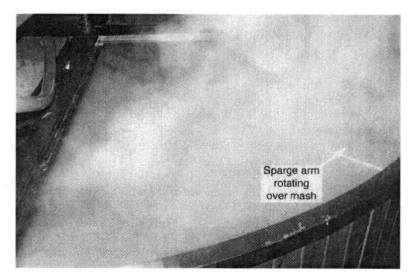

Figure 2.10 *Sparging (Nethergate Brewery)*

until they have run through bright; only then would they be transferred to the copper.

In breweries where there is pressure on the availability of brewhouse equipment (especially the copper), worts may be run into a heated holding vessel, normally situated directly underneath the mash tun. This is the underback.

Over-sparging the mash, usually caused by using sparge liquor of too high a temperature, causes phenols and dextrins to be extracted which are undesirable in sweet wort. This is more likely to happen if the pH of the mash has risen above 5.5.

When the brewer has exhausted all that is feasible from the mash, the grains are drained (Figure 2.11) and removed from the tun, either manually or by being scooped or pumped out. The spent grains have some nutritional value as cattle feed since they contain fibre, cellulose and some protein. If the brewer has done his job properly there should be negligible starch left in the spents.

Mash Filters

The constraints and demands imposed upon the modern brewer have necessitated vigorous attempts to operate at maximum efficiency. This has led to new technology being used in some decoction mash

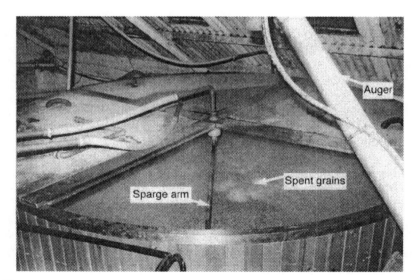

Figure 2.11 *Spent grains after sparging and removal of 'goodness' (Nethergate Brewery)*

breweries – here the lauter tun has been replaced by a mash filter system. New designs of mash filter, notably by Meura/ALC Tournai, are capable of vastly improved efficiency and cost-effectiveness. The new generation of mash filters have polypropylene frames (consisting of a hollow chamber separated by two elastomer membranes) and plates supporting specially woven filter cloths. The whole system is enclosed in a stainless steel case and the plates are closed by a hydraulic cylinder. The process is automatic and there are six phases:

(1) Filling is brought about by the mash being introduced under low pressure through the bottom of the filter. Until filling is complete there is no flow of worts from the filter.
(2) Filtration commences when grains form a layer on the filter cloth.
(3) Precompression; the elastomer membranes are expanded by means of compressed air. This exerts a low pressure on the grains and high gravity wort is extruded. The grains are further compressed onto the filter cloths.
(4) Sparging; sparge liquor is introduced into the bottom filling pipes (where the mash was inserted) and is distributed over the entire grain bed. The rate of speed of sparge liquor introduction has, initially, to balance the release of air from the inflated membrane diaphragms.
(5) Compression occurs when more compressed air is introduced into the system and the membranes are further expanded. This is carried out after sparging is complete. Weak worts are extracted during this phase.
(6) Spent grain removal occurs when the filtration process has terminated and the chambers are opened up. The grains fall into hoppers situated underneath the filter; no manual removal is necessary.

There are numerous advantages to be gained by using these improved wort filters:

(a) The grist can be milled to a much finer degree (one can use a hammer mill) without any detrimental effect on the removal of worts. This will lead to improved extracts. Grists of up to 100% adjunct can be used (barley husks are not required to act as a filter medium).

(b) Yields equivalent to theoretical (or laboratory) values can be obtained.
(c) Worts of high gravity can be produced.
(d) Worts are brilliantly clear with low fatty acid content.
(e) The volume of water used for sparging is greatly reduced.
(f) The whole operation is automatic and rapid, enabling many more brews per day to be carried out.
(g) Spent grains are drier and hence less bulky.
(h) The filter can be easily cleaned by a CIP system (see also page 160).

According to Bamforth,[5] 'the most prominent development in brewhouse technology in recent years has been the Meura mash filter.'

In 1990, Bass Brewers, Burton-upon-Trent, installed two 10.5 tonne mash filters, each having 118 inflation chambers. Each chamber is capable of holding 150 hl of mash (= 85 kg of spent grain) (1 hl = 1 hectolitre = 100 litres). The filters are filled simultaneously *via* bottom ports, the mash being pumped in. This takes four minutes. When the filters are full, filtration commences, and after ten minutes the worts are extremely bright. Filtration lasts 20 minutes and some 235 hl of high gravity worts are extruded and passed to the copper. During the ensuing precompression stage a further 38 hl of strong worts are filtered out and the filter cake is compressed onto the filter sheets. Precompression takes four minutes.

As already mentioned, the rate of introduction of sparge liquor into the filter must balance the rate of air release from the expanded diaphragms. If this is not achieved then clear worts are not produced. Sparging is continued until the required volumes of wort have been recovered. Normal sparge time is one hour and the volume of wort collected during this phase is 225 hl. When sparging has been completed the diaphragms are re-inflated and weak wort is released (about 45 hl). The amount of moisture left in the grains can be controlled by the choice of air pressure used for compression. The normal final compression time is five minutes. At Burton, the moisture content of the spents is standardised at 75%. Spent grain removal takes 15 minutes. Thus, some 545 hl of wort can be taken off the mash in 95 minutes (110 minutes including grain discharge).

Filter cloths have to be cleaned after every 40 brews (the filtration rate gradually slows down), with 3% caustic soda solution being employed at a temperature of 70 °C. The soak time is four hours, with re-circulation of the caustic. The filters are then flushed with acidified water. The whole process takes six hours.

Bass described the advantages of their new mash filter system over their previous lautering operation:

—the turnaround time is two hours;
—wort quality is much higher, especially in terms of brightness;
—worts of 1060° can be produced without having to re-cycle low gravity wort over the grain bed; a process that was necessary in the lauter tun. Worts of 1060° and above require less sparge liquor than would be required by the lautering process (5.3 1 kg^{-1} malt);
—efficiency of extract is much higher than in the lauter tun.

At their Alton plant, Bass installed a new 16 tonne Briggs brewhouse in September 1991 (actually commissioned May 1991). The plant consists of:

—Huppman wet mills (of the short-steep conditioning type, using water at 70 °C);
—a mash mixer (cylindrical with a single agitator);
—a lauter tun (cylindrical with a valley bottom);
—a wort kettle (with external wort-boiling system, or calandria).

The lauter tun was designed to minimise the wort volume under the plates. The mash enters at the side of the tun below mash level.

The whole operation is automatic and initially determined by load cells connected to malt storage bins above the mill. By recognising the load, and the process recipe for the beer being brewed, all other variables are preset until the worts are transferred.

The first, strong, worts are run off at a constant rate and for a specified volume. The bed is then flooded with sparge liquor at a fixed rate and volume.

At the end of the sparge flood, run-off is stopped and the bed is raked. Sparging is then continued, the run-off rate matching the sparge rate. No more raking is necessary.

After sparging, any weak worts remaining in the lauter tun are pumped to weak-wort tanks and will be used for future mashing.

The total run-off time is 130 minutes (the turnaround time is three hours).

Because of the sparge rate (7.5 1 kg^{-1} malt) it is not possible to produce all-malt worts above 1045°. To attain worts above 1045° one has to use weak-wort re-cycling or sugar addition.

CHARACTERISTICS OF SWEET WORT

Carbohydrate Composition

The nature and composition of a sweet wort is entirely dependent upon the mashing protocol and the materials used in the mash tun. The carbohydrate content of a wort will, however, show a similar profile regardless of grist and mashing procedures – as long as industrial enzymes have not been employed in the tun. Some 98% of these carbohydrates come from starch, the remainder (2%) result from the hydrolysis of non-starchy malt substrates such as hemicelluloses. Decoction worts may contain as much as 6% non-starchy hydrolysis products. In most worts, carbohydrates represent some 91% of the wort extract, and anything up to 75% of this fraction will normally be fermentable. Theoretically, after complete hydrolysis, starch yields pure glucose in its monohydrate form (known as dextrose), this is totally fermentable.

There are four main carbohydrate fractions in sweet wort:

(1) *Oligosaccharides* These are principally the dextrins and they represent partial breakdown products of starch. They can constitute between 25 and 27% of total wort carbohydrate. About 80% of wort dextrins possess between 4 and 20 glucose units per molecule, the remaining 20% being formed from over 20 glucose units. Dextrins are non-fermentable and contribute to the calorific value of the finished beer.

The major dextrin is maltotetraose (4 glucose units) but small amounts of maltopentaose, maltohexaose and maltoheptaose are also found. All oligosaccharides are produced during mashing.

(2) *Trisaccharides.* The principal sugar in this group is maltotriose, which comprises some 14% of total wort carbohydrate. There are also trace amounts of iso-maltose. The major proportion of trisaccharides (96%) are produced during mashing.

(3) *Disaccharides.* The main disaccharides are maltose (*ca.* 14% of total wort carbohydrate) and sucrose (*ca.* 5% t.w.c.). Much of the sucrose emanates from malt itself, whilst *ca.* 97% of the maltose present is produced in the mash tun.

(4) *Monosaccharides.* Glucose and fructose are the main mono-saccharides in sweet wort and can comprise some 9–10% of wort carbohydrate. Ribose, arabinose and xylose may also be

encountered in trace amounts. Approximately 16% of these monosaccharides are present in the malt itself, the remaining fraction being formed during mashing.

Also of significance in wort are the β-glucans. They originate from the barley cell walls and become partially solubilised during mashing. The most important, in terms of processing, are the high molecular weight β-glucans (MW >300 000) which, if present, will increase the viscocity of the wort and cause problems during run-off, filtration and clarification. Well-modified malts produce worts with lower molecular weight β-glucans with the resultant fewer production problems. It is now possible to remedy any potential β-glucan problems by incorporating industrial (bacterial) β-glucanases into the mash tun.

Nitrogen Compounds

Nitrogenous constituents in sweet wort consist of an extremely diverse group of compounds, both in terms of molecular nature and molecular size. The main groups of compounds are amino acids, peptides, polypeptides, proteins and nucleic acids. Because of the complexity of the nature of the nitrogenous fraction, they are usually combined, in analytical terms, and referred to as Total Soluble Nitrogen (TSN). Apart from TSN measurement, which is normally *via* the Kjeldahl method, the only meaningful individual nitrogen determination is for free α-amino nitrogen (FAN) which uses ninhydrin colorimetry.

The smaller molecular weight nitrogenous compounds, such as amino acids, are essential for subsequent yeast growth; but the larger compounds are not assimilated by the yeast, and if they persist to the fermentation stage will end up in the final beer. There, they can induce beneficial effects (such as imparting head-retention) or deleterious effects (*e.g.* hazes).

During wort boiling, however, a proportion of the large molecular weight nitrogenous compounds are coagulated and precipitated. The fraction of nitrogenous material left in solution after boiling is referred to as Permanently Soluble Nitrogen (PSN).

Some 50–80% of the amino acids present in sweet wort are derived directly from malt itself. For traditional breweries using an all-malt grist there should be at least 220 mg l^{-1} FAN to support adequate yeast growth.

Trace amounts of *N*-heterocyclic compounds may persist into sweet wort, especially if highly coloured malts have been used in the mash. Most of these compounds are lost during mashing itself or during wort boiling. Many have a very low flavour threshold and the most important *N*-heterocyclics, flavour-wise, are derived from proline.

Amines such as methylamine and ethylamine, which are derived from malt, are often encountered in sweet worts and even finished beer. Enhanced proteolytic activity during mashing will increase their levels. Highly-modified malts produce more amines than under-modified samples. The main significance of amines in wort is their role in the formation of nitrosamines; hordenine, for example, being the precursor of *N*-nitrosodimethylamine (NDMA) (see also section on Nitrosamines in Chapter 6). Compounds such as histamine and tyramine, although they may be present in minute amounts, can cause detrimental physiological effects (*e.g.* headaches).

Fatty Acids

Their presence in wort is totally governed by malt quality. During malting the concentration of fatty acids remains almost constant, but the release of fatty acids into wort, which occurs during mashing, goes hand in hand with protein solubilisation. Fatty acids play an important part in yeast metabolism (membrane synthesis, *etc.*), but too high a concentration leads to staling compounds being formed.

Sulfur Compounds

A wide range of sulfur compounds can be traced back to malt origin, the most important of which are *S*-methyl-methionine (SMM), mercaptans and polysulfides. Most are unstable and possess low boiling points so that many disappear during malt kilning, mashing and wort boiling.

SMM is important because it is the precursor of dimethyl sulfide (DMS) which is a desirable flavour compound in some beers (*e.g.* lagers) and an off-flavour in others (*e.g.* ales). DMS has a boiling point of 37 °C and so is easy to remove during wort boiling. It has a flavour threshold of 70 ppb. It has been shown that the malt curing temperature is all-important in the cleavage of SMM and a significant breakdown will only occur at temperatures above 80 °C (85 °C seems to be optimum). The more modified the malt, the more SMM produced.

OTHER CEREALS USED IN BREWING

Wheat

Wheat (*Triticum aestivum*) is, world-wide, the most extensively grown crop and it is commonly used as an adjunct in the brewing industry, often on the basis of cost. In the continental wheat beers (Weißbier) the majority of the grist may consist of wheat malt, but when used as an adjunct, levels rarely exceed 20% of the grist. This is mainly because the endosperm cell walls contain high levels of pentosans which cause haze problems in the finished beer. This is of no consequence in Weißbiers which are naturally cloudy.

The wheat grain is said to be 'naked', since it possesses no husk. Thus, there are inherent problems during malting because the acrospire is unprotected and can easily be detached from the grain. The lack of a husk, however, does mean that a wheat seed will have approximately 8% more starch per grain, weight for weight, than an equivalent barley seed. In theory, therefore, higher extract levels should be attainable from wheat malt.

Wheat grains are more difficult to mill than barley grains and, for this reason, they are often subjected to a preliminary heat treatment before being used in the brewhouse; this is called torrification or micronisation (see also page 51).

The aleurone layer in wheat is only one cell thick but it responds to gibberellic acid in a similar fashion as barley, except that the α-amylase is produced without necessary stimulation by gibberellic acid.

Unlike some other cereals, the gelatinisation temperature of wheat starch is sufficiently low (52–64 °C) for it to be introduced directly into the mash tun without prior cooking. The lack of a husk, however, renders the grain of limited use in an infusion mash tun.

Rice

Like wheat, rice (*Oryza sativa*) is a naked grain with a single-celled aleurone layer that is responsive to gibberellic acid. The gelatinisation temperature of rice starch, however, is higher than that of barley or wheat starch (70–80 °C) and consequently the grains must be cooked prior to mashing in order to liquefy the starch. World-wide rice production is second only to wheat in tonnage, of which only a small percentage is used in brewing.

Rice grains contain more starch on a percentage dry-weight basis than barley or wheat and they contain lower levels of fibre, lipid and protein, thus possessing some inherently useful properties for the brewer. Being small-grained, rice is low-yielding, in terms of brewer's extract.

The starch structure of rice is more granular than that of barley or wheat.

Oats

The oat seed (*Avena fatua*) contains high levels of lipid and protein and is nowadays rarely used in the brewing industry, although, because of lack of choice, the grain was used extensively during the Second World War. The starch structure is very granular, as it is in rice, but oat starch has a much lower gelatinisation temperature (55–60 °C). The seed possesses a fibrous husk, which is retained during processing, and thus causes no run-off problems in the mash tun.

Rye

Now rarely used in beer production, but employed in the manufacture of certain types of whiskey and bread (especially in Scandanavia). Russia is the largest producer, world-wide. The seed is small-grained, with little or no husk retention and gives highly characteristic flavours to a beer.

Maize

Zea mays, another naked-grained cereal, is used in very limited form in brewing. Again, the aleurone layer is one cell thick, but in maize there is no response to gibberellic acid. The endosperm cell walls are thin and, therefore, β,D-glucan levels are low, which render the starch grains highly available to hydrolysis. Maize starch has a high gelatinisation temperature and so pre-cooking is esential for solubilisation.

Although high in starch content (*ca.* 72%), maize grains also contain a relatively high lipid fraction (4–5%), mainly associated with the embryo. For this reason, before maize can be used for brewing purposes the embryo is removed (the excised embryos being used as a source of oil).

Triticale

The name given to the plant which is a cross between wheat and rye. Considerable experimental work is still being carried out on this plant, which has a higher protein content than barley. Triticale malt has a low fat content, high enzymic activity and high levels of soluble nitrogen. It has only very limited brewhouse use at present. Its seeds are very susceptible to mould growth, especially during germination.

Sorghum

Sorghum vulgare is principally a crop of arid regions and is used, in malted form, to produce a variety of African beers, of which *Kaffir* is probably the most widely known. The seed is again naked and the testa and pericarp contain high levels of polyphenols. There is a single-celled aleurone layer which is not responsive to gibberellic acid. The endosperm cell walls contain high levels of protein which can cause problems in the brewing of non-opaque beers. Starch structure and gelatinisation temperatures are similar to those of maize, but malting losses are very high (can be up to 30%). In addition, the ideal malting temperature for sorghum grains is in the 24–26 °C range; temperatures like these are difficult to maintain in many parts of Africa.

Sorghum is particularly prone to fungal infection during malting and some of the natural surface flora fungi are aflatoxin-producing. Invariably, expensive fungicides have to be employed. Recent work[6] has shown that dilute boric acid/borax solutions used during the early stages of steeping act as an effective antifungal agent and do not impair malt quality.

The lack of response to gibberellic acid and the low level of endosperm cell wall breakdown (due to high protein levels) results in lower (than theoretical) extracts being produced after mashing. There are also low levels of free amino nitrogen (FAN) in the malt which can contribute to poor yeast growth in the brewery.

The sugar content of malted sorghum worts shows a different profile from those of barley and wheat; showing lower levels of maltose and higher levels of glucose.

Because of all of these inherent characteristics, different mashing techniques have to be employed in sorghum brewing.

In 1988, the Nigerian Government imposed a ban on imported barley (and malt), a move which caused severe problems for the lager brewers in that country. The climate in the semi-arid tropics is non-conducive for the cultivation of barley, and so extensive research has recently been carried out on sorghum malting and brewing, most of it aimed at the production of haze-free products (*i.e.* lagers). Traditional

Nigerian sorghum-based fermented beverages, such as otika and burukutu are naturally opaque products.

It has been shown that malted sorghum contains insufficient diastatic (or amylolytic) power to be able to produce a hot water extract (HWE) suitable for brewing purposes. In particular, there is considerably less β-amylase activity in sorghum malt as compared with barley malt and there appears to be no β-amylase activity in the ungerminated grain. Sorghum β-amylase is more thermolabile than the α-amylase and this poses problems during mashing for lager brewing.

The present trend in sorghum brewing seems to be towards the use of raw grain in the mash tun together with industrial enzymes. Alternatively, the mash could consist of 80% raw sorghum, 20% malted sorghum and the relevant addition of enzymes.

Adjuncts

Any materials, other than malt, that are used in the brewing process to produce extract in the mash tun are referred to as adjuncts. In the past, adjuncts have been used as a means of producing extract less expensively than that obtained from malt alone. In some cases this has proved to be a false economy, especially when the additional costs of specialised equipment (*e.g.* cereal cookers) are taken into consideration.

The cereals with high starch gelatinisation temperatures, such as rice, maize and sorghum are often fragmented into small particles called 'grits' before cooking and subsequent introduction into the mash tun.

Some cereals, notably rice and maize, are used in flake form in the mash tun. Flakes are produced by exposing the grains to steam, a process which softens the endosperm. The grains are then passed through rollers to flatten them. Wheat and barley flakes are also occasionally used.

Barley and wheat may also be used in torrified or micronised forms. Torrification involves heating the grains so that they expand and split. This causes partial solubilisation of the starchy endosperm. Micronisation is a similar process, except that the heat imparted originates from infrared rays.

Technically, it is possible to produce acceptable beer with a grist of only 5% malt and 95% adjunct provided that industrially-produced enzymes are incorporated into the mash tun. The highest levels of adjunct usage are found in many US beers where up to 60% maize grits may be incorporated into the mash. In Europe it is uncommon to have beer brewed with more than 40% adjuncts (in the UK it is very uncommon for more than 20% adjunct to be used).

COLOURED MALTS

Malts and barleys with enhanced colours have been used for years in the production of dark beers such as milds, porters and stouts. They also contribute significantly to the flavour profile of a beer. Their production requires specialised equipment and high energy usage. The machines used for their production are modified coffee roasters (roasting cylinders or drums) which are equipped with oil or gas burners, such that a uniform heat is obtained along the drum. Burners can apply either direct heat (*i.e.* through the grain) or indirect heat (around the roasting drum). Drums rotate at about 30 rpm and they are fitted with a series of internal vanes that ensure an even roast.

Coloured malts are prepared from three basic raw materials:

(1) *Dressed barley.* Barley with a moisture level of 12–16% is loaded into the drum and roasted at 230 °C for a period of 2–2.5 hours. When the desired colour has been attained the load is injected with high pressure water which reduces the temperature and causes the grains to swell. The product is Roasted Barley and has a colour in the range of 1200–1500 °EBC (1350° is normal).

(2) *Green malt* (*i.e.* unkilned, well-modified malted barley). This has a moisture level of 42–45% at loading. Direct heat is applied initially for 5–10 minutes in order to remove surface moisture. The drum is then sealed and indirect heating applied for 30–40 minutes. This causes the starchy endosperm to liquefy and saccharify – a process called conversion or mashing. When this is complete, further direct heat is passed through the load to complete the drying and to impart the desired colour (normally 130–135 °EBC after about 2.5 hours). The load is then removed from the roasting drum and placed in coolers, whereby the liquefied endosperm crystallises.

Three categories of coloured malt can be produced by heating green malt, the major one being Crystal Malt (Caramel Malt) which is a widespread constituent of many English ales. Apart from a flavour and colour contribution, crystal malt is said to enhance the shelf-life and head retention of the beer. Crystal malt has a moisture content of 3–5% and a colour in the range of 75–300 °EBC (140 °EBC is normal).

Brown Malt (moisture less than 3.5%) is drier and less sweet than crystal malt and has a lower colour (90–150 °EBC). It used to be produced in kilns heated by hornbeam faggots, which imparted a characteristic smoky flavour. It is used in the production of brown ales and sweet stouts.

Cara Malt (Cara Pils) is characterised by having a low colour (20–35 °EBC), a very glassy endosperm and an enhanced caramel flavour. It is used widely in lagers and low-alcohol beers, where it contributes to foam retention and stability.

(3) *Kilned malt.* Non-cured malt from the kiln is dried to 6% moisture and roasted in a way similar to that of roasted barley; the temperature being lower and certain roasting times shorter. By selecting a heating regime, a range of coloured and flavoured malts can be produced, the major ones being Black, Chocolate and Amber malts. All have a moisture content of less than 3.5%.

Black malt, with a colour of 1200–1400 °EBC, has astringent and smoky characters and is an ingredient of porters and stouts. Chocolate malt has a lighter colour (1000–1200 °EBC) as a consequence of a lower temperature and time of roast. It is less harsh than black malt, but still retains its smokiness. Amber malt is an even more lightly roasted product with soft, dry, baked flavours. The colour is in the range 40–80 °EBC.

The colour specifications for the three malts used at Nethergate Brewery are as follows:

—pale ale (Maris Otter); 4.5–5.5 °EBC;
—crystal; 145–150 °EBC;
—black, 1350–1400 °EBC.

World-wide, there are many varied malt types, most of which are used ultimately to yield beers with specific flavours. Of these, mention may be made of acid malts, which are produced with the aid of naturally-occurring lactobacilli (or the incorporation of lactic acid). Examples are Brumalt, which is made from a long, high temperature steep which eventually becomes anaerobic – yielding a malt with copious quantities of simple sugars, amino acids and enzymic activity, and Munich malt, which is lightly stewed over a period of time to produce a dark colour and high enzymic levels (used for the dark Munich lagers).

COMMERCIAL ENZYMES USED IN BREWING

There are now many commercially-produced enzyme preparations available for use as brewing aids. With a few exceptions, most originate from bacteria and fungi, which are grown in fermenters under controlled conditions. During the overall growth of the microorganism, the enzyme required will be secreted into the culture broth. The supernatant is then drawn off and subjected to partial purification; absolute purification is not achieved because it would yield a product that was inordinately expensive and uneconomical to use. In many cases enzymes other than the one specified may be present, so it is important that some degree of evaluation procedure is carried out before the enzyme is used in a commercial brew. Breweries that employ commercial enzyme preparations must have laboratory-scale evaluation methods, such as the ones developed by the Brewing Research International organisation (BRI). There is always the inherent problem, however, that whatever happens under controlled conditions in the laboratory does not necessarily correlate with the complex events occurring in the brewhouse, particularly in the mash tun. In the final analysis, the brewer has to decide whether the performance of the enzyme in the brewhouse justifies the expense of using it; there must be quality and economic benefits.

All exogenous enzyme preparations must, of course, be of food-grade quality and their activities clearly expressed in units per gramme, or millilitre. In many cases the actual active ingredient required will only form a relatively small percentage of the entire mixture; whether it be in solid or liquid form. Most of the product will be a filler (or diluent for liquid preparations) and it is common to incorporate a preservative and some form of colouring agent (often caramel). The later that enzymes are employed in the brewing process, the more carefully the implications of accompanying compounds in the preparations have to be considered.

The manufacturer's claims for his product should be treated with caution until they can be substantiated by experience in the brewery.

The main groups of enzyme used exogenously in the brewery are:

—glucanases;
—amylases;
—proteases.

(1) Of the glucanase group of enzymes, β-glucanase is probably the most widely used of all commercially-produced enzymes in the brewing industry. β-Glucans in malt can cause a variety of

problems, such as poor extract, poor wort run-off and hazes. The β-glucanase indigenous to barley is temperature-sensitive and may be entirely denatured at 65 °C, a temperature not uncommon in the mash tun. A number of β-glucanases have been isolated which are more heat-tolerant. The main sources are the bacterium *Bacillus subtilis*, and the fungi *Aspergillus niger*, *Penicillium emersonii*, *P. funiculosum* and *Trichoderma reesei*.

When used under practical operating conditions in the mash tun, all products increase wort run-off and reduce wort viscosity, albeit to varying degrees.

Commercial products are assayed by a standard procedure which involves subjecting a purified β-glucan substrate to the enzyme at 30 °C and a pH of 5.0. Activity is expressed in terms of β-glucanase units (BGU); one BGU, by definition, being the amount of enzyme needed to release 1 μmole of reducing sugar. Experience has shown that some preparations are more active under assay conditions than they are in the mash tun.

Different species of micro-organism produce different combinations of enzymes, and what is actually synthesised will be determined by the conditions under which they are grown in the fermenter (especially nature of substrate, pH and temperature). Thus, in theory, microbes which synthesise a variety of glucanases, not just a single β-glucanase, will yield commercial preparations that have greater efficacy in the mash tun. From the brewer's point of view, it is important that the 'degree of impurity' in these products is consistent.

It must be remembered that the breakdown of β-glucans is a very complex process and involves a variety of different enzymes: *endo*-β(1 → 4)-glucanase, *endo*-β(1 → 3)-glucanase, *exo*-β(1 → 4)-glucanase, *etc.*

(2) A variety of starch-degrading enzymes have been used in the brewhouse, many of which can be manufactured with the aid of microbes. Of the major amylolytic enzymes, β-amylase is usually produced from higher plant sources, such as barley, wheat or soyabean, but α-amylase is derived from a number of bacterial and fungal species. Most commercial processes involve the use of the bacteria *Bacillus subtilis* or *B. licheniformis* or the fungus *Aspergillus oryzae*. The action of these enzymes on starch is vastly different, α-amylase from the *Bacillus* spp. operating in an *endo*-fashion and causing liquefaction. High levels of dextrins are produced, but relatively little fermentable sugar. The α-amylase from *A. oryzae*, on the other hand, yields

high levels of maltose and other fermentables (*i.e.* is maltogenic) and in this respect is more akin to the β-amylase of higher plants; its mode of action, of course, is very different.

Glucoamylase, which can be obtained from *Rhizopus niveus, R. delemar* and *Aspergillus niger*, will break starch down to glucose, hydrolysing both the α(1 → 4) bonds and the α(1 → 6) bonds. The related pullulanase enzyme, produced commercially from the bacterium *Klebsiella aerogenes*, debranches the amylopectin fragments of starch since it is only able to cleave α(1 → 6) linkages.

There is sufficient endogenous α- and β-amylase in high quality malted barley to degrade all of the starch present in the grain and so application of pure amylases to the mash tun is usually an unnecessary process. Some preparations, however, do give rise to improved run-off characters and haze levels. It has been shown that this is more attributable to the presence of glucanases as an 'impurity' than to amylase *per se.*

One of the major uses of soybean β-amylase isolate is to increase the maltose level of brewing syrups.

(3) An enormous range of proteolytic enzymes is available commercially, again possessing varied properties. They all act as *endo-enzymes* and, as a result of their activity, amino acids and peptides are produced. Several proteases are isolated from higher plant species and these are generally used as chillproofing agents and are aimed at the degradation of high molecular weight haze precursors. The most widely used are papain, from papaya (*Carica papaya*), bromelain, from the pineapple (*Ananus comosus*) and ficin, from the fig (*Ficus glabrata*).

The proteases derived from microbes fall into two categories according to their temperature optima: the acid proteases which are fungal in origin, and the neutral proteases which can be obtained from fungi and bacteria (*Bacillus* spp.).

Doubt exists as to whether proteases added to the mash tun actually enhance the levels of fermentable sugars in the resulting wort. In theory, they should promote the lysis of the protein sheaths around the starch grains in the endosperm, thereby exposing more grains to amylolytic enzymes. Wort analysis from such mashes suggests that there is an increase in extract, but that this is due to an increase in solubilised nitrogen compounds rather than an increase in fermentable sugar levels.

Protease preparations that have produced better wort run-off rates *in vivo* have been shown to be the ones that contain glucanases as 'impurity'.

REFERENCES

1 A. Ridealgh, *The Brewer*, 1998, **84**, 472.
2 M. Moll, in *Brewing Science, Vol. 1*, ed. J.R.A. Pollock. Academic Press, New York, 1979, p. 539.
3 P. Kolbach, *Wschr. Brau.*, 1941, **58**, 231.
4 D.E. Briggs, J.S. Hough, R. Stevens and T.W. Young, *Malting and Brewing Science, Vol. 1, Malt and Sweet Wort*, Chapman & Hall, London, 2nd edn., 1981.
5 C.W. Bamforth, *The Brewer*, 1996, **82**, 22.
6 K.O. Ajerio, C.D. Booer and M.O. Proudlove, *Ferment*, 1993, **6**, 339.

Chapter 3

Hops

Hop cultivation was first documented in 736AD when the crop was grown at Geisenfeld in the Hallertau region of Bavaria, although there is mention of the plant in Scandanavian sagas dating back to some 1000BC.

The hop was widely grown and used for brewing in Bavaria, Slovenia and Bohemia from the ninth to the twelfth centuries and from these centres of usage its use gradually spread throughout Europe.

The first reference to hops in England is found in a monastic edict of 822AD (by Abbot Abelard of Corvey) which released millers from their duty of grinding malt and hops. Hops were undoubtedly used on a small scale around the time of the Norman Conquest, but the first commercial cultivation of the plant in these islands does not occur until 1524 when Flemish weavers, attracted by the prosperous woollen industry, settled in Kent. They brought their hop technology with them, but there was a general lack of enthusiasm for their utilisation in English ales. The situation was hardly helped by Henry VIII who outlawed the use of hops in the production of 'beer' in favour of the traditional, unhopped, 'English ale'. Edward VI revoked this law in 1552.

The first treatise on English hops was *Perfite Platforme of a Hoppe Garden* by Reynolde Scot (1574), in which details of cultivation were given.

In 1710 the first duty was levied on hops, and the use of other bittering materials was forbidden. By the end of the eighteenth century, hop growing became widespread in Great Britain and some extant varieties, such as English Goldings, were developed. In 1765, hops were being sold at 25 fairs in England and Wales.

By the 1870s the area of land under hop cultivation reached 72 000 acres (29 000 hectares), much of the crop now being grown on wire-work systems instead of the previously-used poles. The first practical wire-work system was devised by Bomford in Evesham in the 1860s. In this country the wire frames are about 5 m high, the hops being grown in rows, with spacings of varying distances depending upon method of harvesting.

Active breeding programmes were instigated at this time, some of which resulted in the production of still much-loved varieties such as Fuggle (developed by Richard Fuggle, 1875).

European settlers to distant parts of the globe took the hop with them, notably to North America and Australia. It is known that hops reached North America in the early seventeenth century *via* English settlers (1629 – Massachusetts Co.), but they did not become a significant commercial crop until 200 years later. Growing was originally confined to New England, close to breweries, but the major growing areas nowadays are further west where climatic conditions are more horticulturally favourable; Washington State, Oregon and Idaho are the centres of production. Similarly, seed and vegetative material reached Australia in the early nineteenth century; the first plantings were in New South Wales, a region not entirely conducive to hop growth. Subsequent, more successful cultivation has been achieved in Victoria and Tasmania.

THE PLANT

The hop commonly used in brewing, *Humulus lupulus*, is botanically placed in the flowering-plant family Cannabinaceae, which, in turn, is in the order Urticales. Thus its botanical relatives include the stinging nettle, elms, bog myrtle and hemp, the latter being its closest relative.

Humulus lupulus is a climbing perennial herb which can reach a height of 6 m. The stems (or bines) climb by twisting in a clockwise direction. The species is dioecious (the male and female organs on separate plants), the most conspicuous morphological difference between the male and female plants being the size of the inflorescence. The male flower head is much branched and some 5 mm in diameter, whilst the female flowers are clustered (often referred to as cones) and of the order of 15–20 mm in diameter. When in fruit the female cone may reach 5 cm in diameter. The flowering period is during July and August and the plant is a native of hedgerows and thickets throughout Europe and Western Asia. Its status in the British Isles was considered doubtful until the 1960s when archaeological work in the Fens yielded pollen and fruit remains that date to around 3000BC.[1] It is not native anywhere in the southern hemisphere although it has been introduced into many countries there.

There are related species found in the Far East: *Humulus japonicus* and *H. yunnanensis*, and there is evidence to suggest that the genus as a whole may have originated in this region. These two species are devoid of resins and are, therefore, of no use in the brewing industry.

In nature, *H. lupulus* is a diploid plant (two sets of chomosomes per cell) with a chromosome number of 20. As a result of breeding programmes, however, triploid varieties exist and some are used in the brewing industry. Although classified as a perennial, it is only the rootstock that has this characteristic since the aerial parts of the plant die back after the flowering season.

To meet the demands of the brewing industry, the hop is now grown throughout the temperate regions of the world, and the species as a whole shows considerable variation as a result. Botanists have attempted to divide the species into a number of sub-species, but, since all forms are interfertile the exercise is somewhat academic.

Because the sexes are separate the species is highly heterozygous, with the resulting extreme morphological variability of plants that are raised from seed. This causes problems for the commercial growers who strive for consistency. For this reason, therefore, a commercial grower will propagate the crop vegetatively, there being three major methods:

(1) from hardwood cuttings;
(2) from growing shoots that have been cut up and planted in sterile peat (called mist propagation);
(3) by 'layering'. This latter process involves the laying down of a growing bine and subsequent overlaying with soil. After the growing season the bine is unearthed, cut up and planted out.

For any form of vegetative propagation, the frequency and position of the leaf nodes relative to the point of cutting is all important. The vegetative reproductive units obtained by any of these three means are called 'setts'.

From a brewer's point of view the useful part of the hop plant is the female cone (or strobilus) which contains all the required compounds. If the cone is dissected the following features are revealed (Figure 3.1);

—a central axis, or strig, bearing a series of nodes;
—at each node on the strig there are two stipular bracts and four

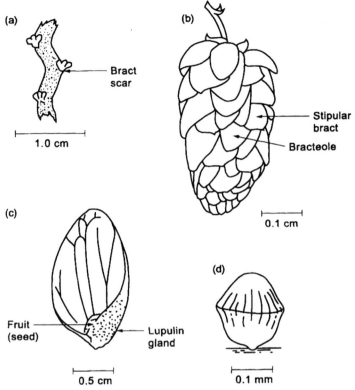

Figure 3.1 *Dissected hop cone* (a) *part of strig;* (b) *mature hop strobilus, or cone;* (c) *isolated bracteole showing lupulin gland;* (d) *dissected lupulin gland*
(Courtesy of Dr D.E. Briggs, Dr R. Stevens and Dr T.W. Young)

bracteoles. The bracts are membranous, persistent and pale yellow-green in colour. The bracteoles have a similar external appearance, but at their bases they bear a single ovary to which is attached a small gland – the lupulin gland. The ovary and gland are situated in an axillary position between bracteole and strig. All of the resins and essential oils (*i.e.* sources of bitterness and aroma) required by the brewer are located in the lupulin glands.

Should the female flower become fertilised by pollen from a male plant then a single seed will develop. This will be enveloped in a tough pericarp, the outer layer of which is membranous and resin-rich.

These resins are not present in unseeded plants and, consequently, the potential bitterness from virgin hops will be lower than that from seeded samples. In spite of this fact, some brewers, especially continental, lager-style brewers, will only use unseeded hops. This is presumably because unseeded samples have a higher essential oil (aroma) to resin (bitterness) ratio and therefore are preferable if predominantly aromatic characteristics need to be imparted to a beer. Some continental brewers also maintain that 'off-flavours' can be produced from the higher levels of fats and waxes found in seeded hops.

Apart from their use as pollinating agents in the production of seeded hops, male hops have no use for the brewer. Besides having relatively few resin glands, the inflorescences are deciduous after flowering. The hop is anemophilous (wind-pollinated) and for the production of seeded hops it is the normal practice to grow one or two male plants per hopyard. Evidence suggests, however, that the majority of female cones are pollinated by plants growing wild in the adjacent countryside. Work carried out early this century indicated that seeded hops gave a greater yield of cones and hence higher resin content per plant. This led directly to the British practice of growing male hops for fertilisation purposes and to the preference for seeded hops by British brewers.

Cultivation of male hops in Great Britain ceased in 1976 when the English Hop Marketing Board, in an effort to expand the export market, adopted a 'seedless hop' policy. The reason for this may be appreciated when it is realised that, in 1950, England, West Germany and the USA all had the same acreage planted out to hops (some 10 000 hectares). By 1975, England lagged far behind. The industry over here was very introspective and was producing hops that were unwanted for continental-style beers. Much of the complacency was attributable to the protective nature of the Hop Marketing Board. There was a danger of the English hop industry collapsing altogether.

It is now evident that some of the newer, high resin (high α-acid) varieties, such as Target, Northdown and Challenger, have higher bitterness potential in the seedless form than when seeded. The converse is true of the older, more traditional varieties such as Fuggle, Goldings and Bramling Cross.

The first commercial seedless hop-growing trials were conducted in Hampshire, a fairly isolated area in terms of established hop cultivation.

The climatological and pediological characteristics of certain areas of the West Midlands and South East England have been largely responsible for the concentration of the British hop-growing industry in these two areas. Like most other plants, the most important single factor governing the ability of the hop to grow successfully in an area is day length and its effect on flowering. The minimum day length for most varieties is 13 hours. Under this regime there will be sufficient vegetative growth to facilitate flower initiation. A shortening of daylight hours is then necessary to permit initiation and subsequent flowering to occur. From a commercial point of view, it is sometimes necessary to delay flowering until economically viable flowering (potential cone) sites have been formed. In countries where the latitude produces non-ideal daylight lengths, artificial lighting is necessary to produce sufficient vegetative growth and/or delaying of flowering. Cold winters are also necessary for the breaking of post-harvest dormancy of the rootstock.

In 1995 the area of land under hop cultivation was as follows:[2]

Kent	1420 hectares
Herefordshire	1080 hectares
Worcestershire	351 hectares
Sussex	149 hectares
Hampshire	59 hectares
Oxfordshire	24 hectares
Surrey	5 hectares

One of the very significant developments in hop cultivation was 'low-trellis' growing which was pioneered in Yakima, Washington State, USA by the German company Hop Union in 1985. At first, English growers shunned such a practice on the grounds that high winds, especially in August, would cause problems for a trellis laden with a heavy crop. With the development of a relatively inexpensive harvester, however, the industry over here has now accepted the concept of low-trellis growing for traditional varieties – as long as the supporting poles are sunk to a depth of at least 4 ft. The first British low-trellis hop gardens were set up in 1992. It has been calculated that the cost of setting up a low-trellis garden is some 30% less expensive than for a traditional frame garden, whilst concomitant labour costs are around 40% lower.

Over the last century or so there has been a gradual diminution in the acreage given over to hop cultivation in the UK, this being directly related to the per capita fall in the volume of beer drunk in these islands. As a consequence, from the highest recorded area under hop cultivation (29 000 hectares in 1870) only 3088 hectares were planted in 1995.

In a somewhat gloomy résumé of the hop industry, Ibbotson[3] reports that world hop production has decreased by 25% since 1996, with a 15% reduction in the US and a 5% reduction in Germany (who between them are responsible for some 65% of world production). The reduction in the US crop is partly attributable to a serious, widespread outbreak of powdery mildew, particularly in Washington state. With the modern brewer's predilection for high α-acid, the Americans with their 'super alpha' varieties (at least 16% α-acid) have recently dominated the world hop market. Ibbotson envisages the demise of UK varieties such as Challenger; with only Target being able to survive in the long term. As he says, 'beer production is, at best, static', and this, coupled with the fact that brewers now use hops more efficiently, will lead to many growers seriously pondering the future.

PROCESSING

After hops have been harvested, by whatever means, they have to be dried before they can be stored. Drying traditionally takes place in kilns or oast houses, where hot air is blown over the crop under controlled conditions; air speed and temperature are the all-important criteria. Drying must be carried out as soon as possible after harvest. At the point of harvesting, hops have a moisture content of 75–80% and this must be reduced to 10% to enable storage for any period of time. The main problem to be encountered during the drying phase is overheating, which causes discolouration and a loss of α-acid content. To overcome the former, it was standard practice, until recently, to burn sulfur in the kiln. This gave the hop sample an overall, standard yellowish tinge which was favoured by some brewers in their hand-evaluation of the crop. Because of the problem of residues, the burning of sulfur in the kiln has now been abandoned.

In the UK, after drying, hops in their whole form are packed into sacks known as pockets. Traditionally these were made of jute; nowadays they are made of polypropylene. A pocket contains 1.5 cwt. (76 kg) of hops and is some 7 ft long by 2 ft in diameter. Before 1850, hops were trodden into pockets, but in that year the first hop press was invented by Ellis Hopgood of Barming in Kent.

In the USA, hops are packed into bales containing 200 lb (90.7 kg) of product. These are rectangular in shape.

By definition, whole hops deteriorate on storage, especially in terms of aroma and resin loss. This propensity can be reduced by low-temperature storage (4 °C). Some varieties deteriorate more rapidly than others.

The standard commercial weight unit for whole hops is the Zentner, which is equivalent to 50 kg.

For a thorough treatise on the hop, the book by Neve[4] is *par excellence.*

HOP PRODUCTS

The pocket, or bale, represents a cumbersome way to store and transport hops, especially in light of the fact that there is only really 5–15% active material in that bulk for the brewer. To counteract the problem of bulk, and the large volume to weight ratio, the volume of hops can be reduced by compression. This will give a smaller pocket or bale size; approximately half-size. Before hops can be compressed they must have less than 10% moisture content.

Over the years, other processing methods have been developed, all of which result in more manageable ways of storing and transporting hops. The main processed products are hop powders, hop pellets and hop extracts.

Hop powders are comminuted products produced by carefully controlled milling of pure, dry (6–7% moisture) hop cones. Drying is effected by heating the cones to 65 °C, after which the temperature is reduced to around −35 °C to make the hops less sticky. Milling occurs at this lower temperature. The resultant powder requires about one-sixth of the storage space of the whole cones themselves. Powders can be enriched by low-temperature (−35 °C) sieving, which, in essence, separates the lupulin glands from the coarser material. Up to 50% concentration of the hop goodness can be achieved by this method. The storage of hop powders has always been problematical, particularly because of the ease of loss of resin and aroma characters (owing to the high surface area of the particles). In 1962, vacuum-packing was introduced which served to preserve the essential ingredients to some degree. In spite of this, hop powders are usually converted into pellets which, with a lower surface to volume ratio, suffer less from oxidation problems.

Hop pellets exist in several forms, the standard one being the Type '90'. This is produced by re-drying the hop sample to 7% moisture and then milling, homogenising, pelletising, cooling and vacuum-packing. Related to the Type '90' is the Type '100' pellet in which the hops have not been re-dried before milling.

Pellets can be enriched by mechanical separation of lupulin glands such that enhanced levels of α-acid are imparted. The principal enriched pellet is the Type '45', but this exists in a more concentrated form (Type '33') and a less concentrated form (Type '75'). In spite of their obvious advantages, hop pellets give rather poor utilisation rates in the brewery. For example, the conversion of potential bittering (α- and β-acids) to actual bittering (the isomerised forms) rarely exceeds 45% during a one hour boil. Several attempts were made either to protect the α- and β-acids from oxidation or to improve the isomerisation during the boil.

In 1978, Grant[5] found that addition of between 1 and 3% magnesium oxide to the hops during pelletisation significantly increased the isomerisation rate during boiling. This is attributable to the fact that the magnesium salts of the α- and β-acids go into solution far more rapidly and are isomerised more readily. These are called stabilised pellets. Bentonite has also been incorporated into the pelletisation process; the resulting pellets give 20% better utilisation during boiling. This is due to the fact that the bentonite powder provides a larger surface area for isomerisation.

In 1979, Grant[6] discovered that by heating his stabilised pellets to 80 °C for two hours (in the absence of oxygen), isomerisation of the α- and β-acids occurred *in situ*. These pellets were even more stable than their predecessors and did not need to be stored at low temperature. These are called isomerised pellets, and high levels of utilisation (60%) are attainable with them. An additional advantage is that shorter boiling times are possible.

The first attempts to extract and isolate the active ingredients of hops, with a view to improving their storage characters, were made last century, with water being the initial solvent. Since then a variety of solvents have been used, with varying success. They include ethanol, dichloromethane, trichloroethylene, benzene, methanol, hexane, methylene chloride and liquid carbon dioxide. It is only really since the 1960s that hop extracts have been commercially produced and used on any significant scale, and it is now estimated that some 50% of world hop harvest is used in the form of extracts. One of the major problems encountered in the production of hop extracts has been the loss, through volatilisation or chemical rearrangement, of some important constituents. The degree of loss is dependent upon the solvent used and the extraction system.

With modern food regulations it is likely that only ethanol and liquid carbon dioxide will be used to any great extent, because there is a trend away from the use of 'unnatural' ingredients in food and beverage processing. In 1994 an advanced ethanol extraction system was developed in Mainburg, Germany, whereby all of the desired hop constituents are extracted and recovered. By improving the vacuum in the evaporation plant it has been possible to reduce the processing temperature, thus enabling even the most volatile of hop constituents to be recovered. To the delight of environmentalists, the ethanol is produced from the fermentation of plant-based raw materials. Using this method, resin extracts have been produced which are tannin-free.

Extracted hops can be presented in a variety of forms, according to how the brewer wishes to use them in his plant.

(1) *Conventional or kettle extracts.* Probably the most widely used form at present, although the situation is changing. α-Acids and all other constituents are extracted in varying ratios and at varying efficiencies; ideally at least 95% extraction rate. These extracts are added to the copper (kettle) in the same way as whole leaf hops would be. It is claimed that the brewer can produce a far more consistent product with extracts; certainly their storage and stability characters are advantageous. More recently, pre-isomerised kettle extracts (PIKE) have been developed which can enable boiling times in the brewhouse to be reduced. These give greatly enhanced α-acid utilisation rates.

(2) *Isomerised extracts.* These are derived from liquid carbon dioxide extracts. Isomerisation is either by heating with aqueous alkali, which produces the corresponding salt (usually potassium), or by converting the acids to calcium or magnesium salts and subsequently heating. The latter process yields calcium or magnesium salts in the form of a suspension or a powder. Isomerised extracts are used for post-fermentation bittering purposes and they represent a very efficient means of using (and controlling) α-acid in a beer.

(3) *Speciality extracts.* The essential oil content of conventional and isomerised extracts is normally in the range 3–8% vol/wt, but this can be increased by a fractionation process devised by Sharpe *et al.* at the Brewing Research Foundation (now Brewing Research International) in 1980. By passing liquid carbon dioxide through a column of powdered hops, they found that the first eluted fraction contained principally essential oils (78% extraction in a 30 minute running time). Such aroma extracts are used to impart hop flavours to cask beers and are usually added to the cask itself (*i.e.* a form of dry-hopping). By adjusting the temperature of extraction, the BRF team were able to produce fractions containing almost entirely essential oils: α-acid extract and β-acid extract.

It is well known that iso-α-acids (and their β-counterparts) are very sensitive to light, especially in the 350–500 nm range. Light of this

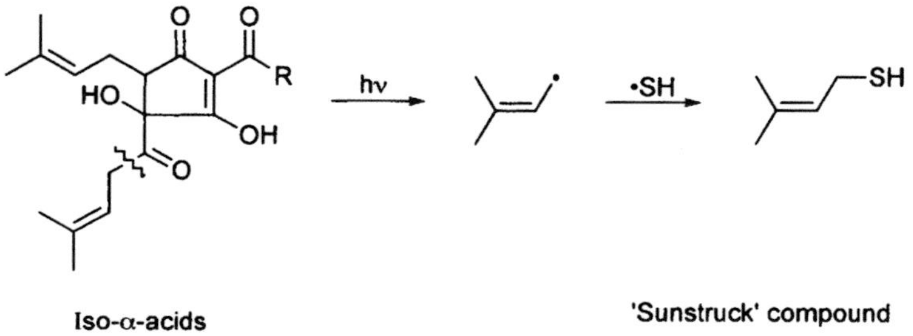

Iso-α-acids   'Sunstruck' compound

Figure 3.2   *The light-struck reaction*
     (Courtesy of the International Brewers' Guild)

wavelength can penetrate clear and green glass and cause nauseous off-flavours in beers bottled in such glass containers. The beer is said to be 'sunstruck' and the taste referred to as 'skunky'. What happens is that one of the side chains on the iso-α-acid is cleaved and the highly reactive radical that is liberated combines with sulfur-containing compounds (see Figure 3.2) to produce 3-methyl-2-butene-1-thiol (MBT). MBT has a flavour threshold in the order of parts per trillion, making it one of the most flavour-active substances in beer. Reduced isomerised hop extracts (RIHE) have now been developed to combat this sensitivity to light. In essence, pairs of hydrogen atoms are catalytically added to the isomerised α-acid, with the resultant production of dihydro-, tetrahydro- and hexahydroiso-α-acids. These extracts are not light-sensitive and the last two have also been shown to be beneficial to foam stability.

A fine review of the manufacture and uses of hop products has been written by Combes.[7]

HOP VARIETIES

Hops are graded into three categories, according to their required function in the brewhouse: (a) aroma hops, (b) alpha hops and (c) dual purpose hops. The terms are self-explanatory.

Table 3.1 shows a list of some important varieties that have been used over the years, with their country of origin and inherent characteristics.

The early hop growers, world-wide, would certainly have used wild plants to intitiate their stocks, and, through a process of elimination, produced clones which were more vigorous and productive. Natural selection would obviously have played a part, but there is evidence that selective breeding was carried out during the early years of hop usage. Improved travel facilities resulted in the transport of proven

Table 3.1 *Hop gradings and varieties*

Variety	Country of origin	Percentage composition		
		α-Acid	β-Acid	Essential oil
(a) Aroma hops				
Goldings	England	5.5	2.5	0.7
Fuggle	England	4.5	3.7	0.6
Progress	England	6.0	2.3	0.6
Whitbread Golding Variety	England	6.0	2.7	1.0
Hersbrucker	Germany	4.2	7.0	0.8
Tettnang	Germany	4.0	6.0	0.8
Spalt	Germany	4.0	7.0	0.8
Hallertau	Germany	4.5	6.0	0.8
Willamette	USA	6.0	4.0	1.0–1.5
Cascade	USA	6.0	5.0	1.0
Mount Hood	USA	5.0	4.0	1.0
Strisselspalt	France	4.5	2.0	0.7
British Columbian Bramling	Canada	5.0	2.3	0.7
Styrian Goldings	Slovenia	5.0	2.6	0.8
Saaz	Czech Republic	3.0	7.0	1.0
Lublin	Poland	5.0	1.3	1.0
N.Z. Hallertau	New Zealand	7.5	6.0	1.0
(b) Alpha hops				
Target	England	10.5	5.2	1.3
Yeoman	England	10.5	5.1	0.8
Omega	England	8.0	3.5	1.0
Northern Brewer	Germany	8.0	7.0	1.6
Brewers' Gold	Germany	6.5	6.5	1.4
Rekord	Belgium	6.0	6.0	1.0
Galena	USA	13.0	8.0	1.2
Nugget	USA	13.0	5.0	2.0
Clusters	USA	7.0	5.0	0.5
Pride of Ringwood	Australia	9.0	1.7	2.0
Super Styrian	Slovenia	8.5	2.2	1.0
Green Bullet	New Zealand	10.0	7.8	1.3
Pacific Gem	New Zealand	12.0	9.2	1.8
(c) Dual purpose hops				
Northdown (seeded)	England	8.0	6.0	1.0
Northdown (seedless)	England	10.0	6.6	2.0
Challenger (seeded)	England	7.0	3.9	0.8
Challenger (seedless)	England	9.0	4.5	1.3
Perle	Germany	6.5	8.0	1.0
Hüller	Germany	5.5	6.0	1.0
Centennial	USA	10.0	5.0	1.5
Chinook	USA	13.0	3.0	2.0

stock material being distributed widely throughout beer-drinking countries. In the early part of this century a vast number of hop varieties were grown world-wide, but by the 1940s the demands of brewers had resulted in a huge reduction in number, such that most major hop-growing regions were only cultivating one or two varieties. Examples of this situation include Fuggle (in England), Clusters (USA), Saaz (Czech Republic) and Hallertau (Bavaria).

The inauguration in 1894 of Wye College, Ashford, Kent saw the first scientific investigations into hops. The succeeding hundred years have seen the undertaking of an immense amount of research, including the implementation of several important breeding programmes. Some of this work has been carried out in conjunction with the nearby East Malling Research Station, which merged with Wye College to form part of the Institute of Horticultural Research, now known as Horticultural Research International.

One of the earliest programmes at Wye was conducted by Prof. E.S. Salmon and was aimed at counteracting the import of American high α-acid Clusters hops, which British brewers had started to demand. Salmon's programme resulted in the production of three new varieties: Brewers' Gold, Bullion and Northern Brewer. All of these had higher α-acid content than anything else available in England at the time. For a number of reasons these three varieties have subsequently been more successful on the continent than in the UK.

In the 1970s, Dr Ray Neve, again of Wye College, bred a series of high α-acid varieties which have found more general acceptance in the UK. These were Wye Challenger, Wye Northdown and Wye Target, the latter not only having a very high α-acid level but also being resistant to wilt. During the three preceding decades, W.G. Keyworth had carried out prolific work on wilt-resistant varieties at East Malling, particularly with a view to finding a substitute for the very wilt-prone Fuggle variety. One such successful alternative was Bramling Cross. Independent work by Whitbread & Co. at Beltring, Kent, produced the most successful of all English wilt-resistant aroma hops, WGV (Whitbread Golding Variety), which is still much in demand today.

Continental, especially German and Czechoslovakian, hop breeding has traditionally been aimed at the production of enhanced aroma varieties rather than high α-acid varieties. The spread of downy mildew and wilt throughout Europe led to the foundation of a research station at Hüll, in Germany, in 1926. A variety called Hüller Bitterer was developed which was resistant to both diseases. This variety was superseded by Perle, which proved even more successful in terms of brewing acceptability.

As in other forms of agriculture, the search for new, improved varieties is never-ending and most major brewing countries now have hop research establishments.

Dwarf Hops

Probably the most important recent development in the hop industry has been the production of dwarf varieties. In 1977, Ray Neve, perchance noticed an unusual plant in the breeding nursery at Wye College farm. Its main characteristic was that it had shortened internodes on the bine. The plant proved to be fundamental to an extensive breeding programme which resulted in accepted (to the brewer) dwarf varieties being involved in farm trials in 1993. The original plant had an α-acid content of 1%, but after hybridisation and selection this had risen to 13% in the variety called Herald. Dwarf plants now have acceptable aroma and storage characteristics, but there are still a few shortcomings to be counteracted.

The first generally accepted dwarf hop was the variety First Gold, which was introduced in 1995 and resulted from a cross pollination of WGV with a dwarf male. Used extensively in 1996, it proved to suitable as a kettle hop and for late and dry-hopping. It has a well-balanced bitterness and excellent aroma qualities. In 1996, the average α-acid content was 9%, whilst β-acids were 3.0–4.1% and total oils 0.7–1.3%. First Gold is susceptible to downy mildew, resistant to powdery mildew, and is reasonably tolerant to wilt.

Dwarf varieties are grown on 2.44 m (7.87 ft) high wire-work in individual rows, the plants themselves growing to a natural height of 3 m. A crop of this stature lends itself well to mechanical harvesting and the consequent labour savings. The crop is also easier to spray with biocides, easier to inspect for diseases and less expensive to set up, garden-wise.

On harvesting a dwarf hop the bines are left intact (in a traditional harvest the bines and string framework are cut down as well), which means that nutrients continue to be passed to the rootstock after harvest. This gives such rootstocks more vigour for growth in the following season. It is forecast that dwarf hops will be a major feature of the industry by the twenty-first century.

HOP CONSTITUENTS

In general terms, a good commercial hop sample is expected to have the following composition:

Component	% of sample
Water	10.0
Total resins	15.0
Essential oils	0.5
Tannins	4.0
Pectin	2.0
Lipids and waxes	3.0
Monosaccharides	2.0
Proteins ($N \times 6.25$)	15.0
Amino acids	0.1
Cellulose and lignin	40.4
Ash content	8.0

Resins are classified on the basis of their solubility in various organic solvents. The total resin content of a hop is that fraction which is soluble in both cold methanol and diethyl ether. Extraction of hops with diethyl ether will dissolve resins, waxes and fats, but only the resins are soluble in methanol. If the total resin fraction is dissipated in hexane, only some of them actually dissolve; these are the soft resins. The hexane-insoluble fraction is referred to as the hard resins.

Most of the inherent bitterness of the hop is contained in the soft resin fraction, which itself can be divided into two main categories according to the reaction with lead acetate solution. If a methanolic solution of lead acetate is added to a methanolic solution of soft resins, a yellow precipitate forms; this represents the α-acid fraction. The supernatant represents the β-acid fraction. The formation of an insoluble lead salt provides the basis for the standard method of measuring α-acids in hops. This is the IOB-approved conductimetric method by which a Lead Conductance Value (LCV) is obtained (IOB Method 6.4, 1997).

Some compounds in the soft resin component are not attributable to either α- or β-acid fractions; these are the uncharacterised soft resins.

The main α-acids are humulone, cohumulone and adhumulone, whilst the most important β-acids are lupulone, colupulone and adlupulone. As can be seen from Table 3.2, these compounds differ only by the nature of their acyl side chains.

The first soft resin component to be isolated was lupulone[8] last-century, but its structure was not elucidated until 1925 after extensive work by Wollmer[9] and Wieland[10,11] over a period of ten years. The same workers also derived a structure for humulone, which had been

Table 3.2. *Analogues of the α- and β-acids*[†]

α-Acids

β-Acids

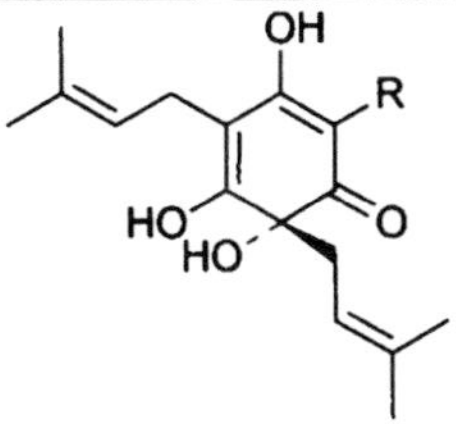

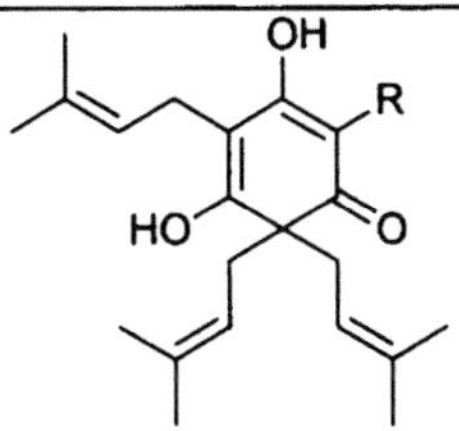

Acyl side chain (R)	α-Acids			$[\alpha]_D^{26}$ (°)	pK_a	β-Acids		
	Name	Formula	Mp (°C)			Name	Formula	Mp (°C)
–COCH₂CH(CH₃)₂ isovaleryl	Humulone	$C_{21}H_{30}O_5$	64.5	–211	5.5	Lupulone	$C_{26}H_{38}O_4$	92
–COCH(CH₃)₂ isobutyryl	Cohumulone	$C_{20}H_{28}O_5$	Oil	–208.5	4.7	Colupulone	$C_{25}H_{36}O_4$	93–94
–COCH(CH₃)CH₂CH₃ 2-methylbutyryl	Adhumulone	$C_{21}H_{30}O_5$	Oil	–187	5.7	Adlupulone	$C_{26}H_{38}O_4$	82–83
–COCH₂CH₃ propionyl	Posthumulone	$C_{19}H_{26}O_5$					$C_{24}H_{34}O_4$	101
–COCH₂CH₂CH(CH₃)₂ 4-methylpentanoyl	Prehumulone	$C_{22}H_{32}O_5$	Oil	–172			$C_{27}H_{40}O_4$	91
–COCH₂CH₂CH(CH₃)CH₂CH₃ 4-methylhexanoyl							$C_{28}H_{43}O_4$	91

[†] Courtesy of the Institute of Brewing.

isolated by Lintner and Schnell in 1904.[12] Lupulone and humulone were the only soft resin compounds known until the 1950s, when counter-current distribution techniques enabled Rigby and a series of co-workers to determine that α-acids were a mixture of homologues and analogues.[13] Their work resulted in the identification of cohumulone and adhumulone, and further studies by Verzele in 1958[14] led to the separation of prehumulone and posthumulone.

Humulone, the major α-acid, is a pale yellow solid (mp 70–72 °C), highly soluble in all organic solvents, fairly acidic (pK_a 5.4), and has a faint bitter taste. Weiland's original structure, derived in 1925, was slightly modified by Verzele and De Poorter in 1959[15], and finally by De Keukeleire and Verzele in 1970,[16] who proposed the enol form (see Figure 3.3).

The percentages of the individual α-acids found in hops are shown in Table 3.3.

Separation of the β-acids using counter-current distribution proved

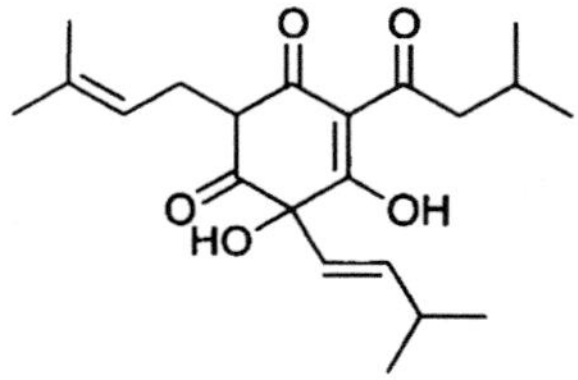

(a) original proposal by Weiland (1925) as a result of alkaline hydrolysis and hydrogenation

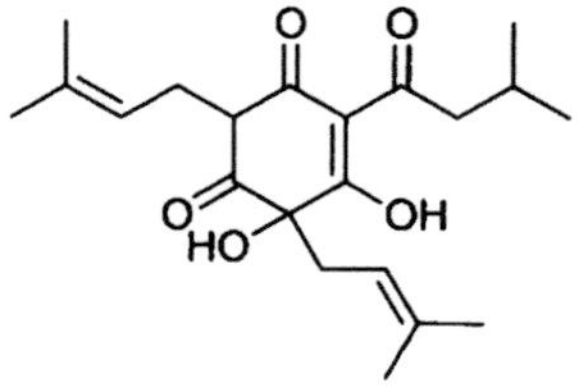

(b) proposal by Verzele and De Poorter (1959) as a result of ozonolysis experiments

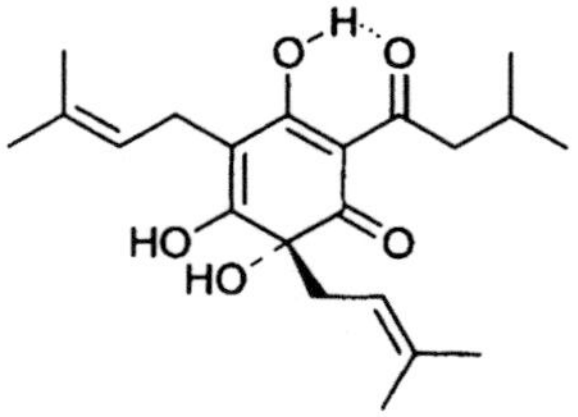

(c) the final enol form devised by De Keukeleire and Verzele (1970) from ORD and CD spectra

Figure 3.3   *The structure of humulone*
(Taken from *Brewing Science, Vol. 1*, ed. J.R.A. Pollock.
© (1979) Academic Press, UK)

Table 3.3 α-*Acids found in hops*

α-*Acid*	*Percentage found*
Humulone	35–70
Cohumulone	20–55
Adhumulone	10–15
Prehumulone	1–10
Posthumulone	1–5

more difficult because of their extreme sensitivity to oxygen. Early work by Rigby in 1951[17] and by Howard and co-workers in 1955[18] did provide some means of characterisation, but it was not until Verzele used HPLC in 1979[19] that a satisfactory method of determining β-acids was developed.

The first synthesis of soft resin components was carried out by Riedl in the early 1950s, with humulone being synthesised in 1951. In a series of subsequent publications by Reidl (in 1952,[20] 1954[21,22] and 1958[23]) he reported the synthesis of a number of α- and β-acids and their analogues. The synthetic soft resin compounds were all racemic, whereas those that are naturally hop-derived are optically active. All of the natural α- and β-acids have now been synthesised.

The core compound for all hop resins is phloroglucinol (1), and the first stage of synthesis is acylation to phloracylphenone. Subsequent alkylation with isoprene hydrobromide yields a range of isoprenylated derivatives.

Phloroglucinol

(1)

Essential oils, or hop oils, are, like the resinous compounds, mainly derived from the lupulin gland. They contribute to the aroma, or bouquet, of a beer and in some instances are solely responsible for the beer style. For any given variety of hop, seedless forms always contain more essential oils than the seeded forms (the reason for continental brewers' preference). Maximum biosynthesis of essential oils occurs during the later stages of hop maturation, after the majority of resin synthesis has been completed.

The essential oil fraction can be divided into three component groups, one of which can be distinguished from the other two by the fact that it can be eluted with light petroleum from silica gel columns or plates. This is the hydrocarbon fraction, and comprises 50–80% of the total essential oil content. The other two fractions consist of chemically-bound oxygen compounds (20–50%) and chemically-bound sulfur compounds (less than 1%). Over 300 compounds have now been separated from hop oils by modern chemical methods.

The major hydrocarbons are monoterpenoid and sesquiterpenoid in nature, and about 40 have been characterised from hop oils, although relatively few are known to be present in all hop varieties. The most widely distributed, and abundant, in terms of quantity, are the monoterpene myrcene (C10), and the sesquiterpenes humulene and caryophyllene (C15), as illustrated in structures (2)–(4).

| Myrcene | Humulene | Caryophyllene |
| (2) | (3) | (4) |

Myrcene is generally considered to be an undesirable component of the essential oil portfolio, in spite of its relative abundance.

Very few hydrocarbons can survive an extended wort boiling process in order to influence the flavour of beer. For this reason aroma varieties are often added late to the copper, especially if characteristic floral aromas are required in the final product; this is the case in some continental lagers.

Chemical analysis of hop samples over their ripening period indicates that the synthesis of cyclic terpenes, such as caryophyllene and humulene, is completed first and a definitive level of these compounds is reached. The same is not true for the monoterpenes, myrcene especially, which are synthesised right up to the end of ripening.

It is now accepted that the ratio of caryophyllene to humulene and the abundance of certain other hydrocarbons is variety-specific and, accordingly, modern hop variety specifications document the percentage quantity of these marker substances.

The specifications for the three varieties used at Nethergate, namely Fuggle, Challenger and Styrian Goldings, are illustrated in Table 3.4.

Over 60 compounds containing chemically-bound oxygen have been identified from hop oil and, as a group of chemicals, are very diverse,

Table 3.4 *Résumé of three hop varieties:* (a) *Fuggle,* (b) *Challenger,* (c) *Styrian Goldings*

(a) Fuggle	
Class	Aroma

Origin and production
The Fuggle was propagated in Kent by Mr Richard Fuggle in 1875. It became the most widely grown hop in England until Verticillium Wilt made growth almost impossible in much of Kent and Sussex. It is still grown in the West Midlands and in the USA, mainly in Oregon.

Analytical data (typical)

Alpha acid (range)	4.0–5.5%	Ratio alpha : beta
(IOB, LCV)		1.8 : 1
Beta acid (range)	2.2–3.1%	
Cohumulone (% of alpha)	26	
Total oils [ml 100 g^{-1} (dry)]	0.7–1.1	
		Ratio to
Caryophyllene	12.4%	caryophyllene
Myrcene	25.6%	2.1
Farnesene	5.8%	0.5
Humulene	37.1%	3.0
Selinene	1.7%	0.1

Plant characteristics

Yield Ztr/ha (average)	28
Time of ripening	Early/mid season
Wilt sensitive/tolerant	Sensitive
Other diseases	Susceptible to powdery mildew; some resistance to downy mildew
Storage stability	Good

Brewing uses

Pellets,	Yes
Extracts	No
Kettle hop	Yes
Dry-hopping	Yes
Essential oil/emulsion	Sometimes

Flavour
Typical English, frequently blended with Goldings to improve 'drinkability', adding roundness and fullness to the palate. Oregon Fuggles tend to have lower caryophyllene and humulene contents, with higher myrcene; brewing quality is unchanged.

Table 3.4 *Continued*

	(b) Challenger
Class	Dual purpose

Origin and production
Challenger was bred at Wye College and released for commercial production in 1972. It has good growing characteristics with heavy yields and a fine English-type aroma. It is grown in England in the West Midlands and Kent, and also in small quantities in Belgium and France.

Analytical data (typical)

Alpha acid (range)	6.5–8.5%	Ratio alpha : beta
(IOB, LCV)		2.0 : 1
Beta acid (range)	3.2–4.2%	
Cohumulone (% of alpha)	22	
Total oils [ml $100 \, g^{-1}$ (dry)]	1.0–1.5	
		Ratio to
Caryophyllene	9.4%	caryophyllene
Myrcene	30.8%	3.3
Farnesene	1.6%	0.2
Humulene	29.2%	3.1
Selinene	12.3%	1.3

Plant characteristics

Yield Ztr/ha (average)	36
Time of ripening	Late season
Wilt sensitive/tolerant	Sensitive
Other diseases	Some resistance to downy mildew; susceptible to powdery mildew
Storage stability	Good

Brewing uses

Pellets	Yes
Extracts	Yes
Kettle hop	Yes
Dry-hopping	Yes
Essential oil/emulsion	Yes

Flavour
Fruity, almost scented, and blends well with other English varieties. A versatile kettle hop for all types of beer; also used as late hop and dry hop.

Table 3.4 *Continued*

(c) Styrian Goldings	
Class	Aroma

Origin and production
Grown in Slovenia, mainly in the Savinja Valley around Zalec, this variety is genetically the same as Fuggle but has developed distinctive flavour characteristics peculiar to that environment.

Analytical data (typical)

Alpha acid (range)	4.0–6.0%	Ratio alpha : beta
(IOB, LCV)		2.0 : 1
Beta acid (range)	2.0–3.0%	
Cohumulone (% of alpha)	28	
Total oils [ml 100 g^{-1} (dry)]	0.5–1.0	
		Ratio to
Caryophyllene	10.1%	caryophyllene
Myrcene	29.8%	2.9
Farnesene	3.4%	0.3
Humulene	36.2%	3.6
Selinene	1.5%	0.1

Plant characteristics

Yield Ztr/ha (average)	30
Time of ripening	Early/mid season
Wilt sensitive/tolerant	Sensitive
Other diseases	Susceptible to downy and powdery mildew
Storage stability	Good

Brewing uses

Pellets	Yes
Extracts	Rarely
Kettle hop	Yes
Dry-hopping	Yes
Essential oil/emulsion	Yes

Flavour
A distinctive, mild flavour, somewhat richer than English Goldings. Often used as a late kettle addition and for dry-hopping.

encompassing esters, ketones, epoxides, alcohols, acids and aldehydes. Some of these compounds are extremely volatile, or even transient, and few survive wort boiling. Many are formed as intermediates in the synthesis of other major hop constituents. Esters are numerically the most significant oxygenated compounds and some persist into chilled wort where, during fermentation, transesterification can occur. The levels of these compounds will increase during hop storage, partly because many are derived from oxidised hydrocarbons. It has now been established that there is a generic relationship between hop hydrocarbons and oxygenated derivatives.[19]

Although only present in trace amounts in most hop varieties, the majority of sulfur-containing compounds have very low flavour thresholds and, therefore, contribute greatly to the beer flavour profile. This is especially the case where hops are added late in the brewing process (*i.e.* late copper or post-copper). The first sulfur compound to be reported from hop oil was diallyl sulfide.[24] The invention of flame photometric detection (FPD) and advances in the techniques of gas chromatography coupled with mass spectometry (GC–MS) subsequently led to the detection and identification of many more volatile sulfur compounds. The major groups found thus far are thiols, sulfides, polysulfides, thioesters and thiophenes. In addition, some terpenes (especially humulene and caryophyllene) react with elemental sulfur (as used in the field to counteract mildew or in the kiln during drying) to form episulfides. Over 30 sulfur-containing compounds have now been identified from hop oil.

As a group, the alkyl sulfides and polysulfides are probably the most important in terms of their ability to impart flavour to beer. Diallyl sulfide itself is not now thought to be a natural constituent of hop oil, but dimethyl sulfide (DMS), dimethyl disulfide (DMDS) and dimethyl trisulfide (DMTS) certainly are. DMTS is particularly flavour-potent, having a flavour threshold of 0.1 ppb. It is formed from a very labile precursor, (*S*)-methylcysteine sulfoxide. DMTS is said to impart a rubbery flavour reminiscent of cooked vegetables! Indeed, the UK practice of using sulfur dioxide in hop kilns was abandoned in 1981 because of the production of 'off odours' in beers that were late-hopped.

DMS is regarded as contributing desirable flavours to beers, especially those that have been lagered. As well as being a constituent of hop oil, DMS also originates from a heat-labile precursor (*S*-methylmethionine), present in some malt types.

HOP DISEASES

These are caused by certain fungi, arthropods and viruses, although arthropod-caused diseases are not of serious consequence in the UK.

Verticillium Wilt

Verticillium wilt, caused by *Verticillium albo-atrum*, was first reported in Kent in 1923 although it is now recognised in most hop-growing countries. In 1930 a far more virulent strain was encountered in a Fuggle garden, also in Kent. This strain spread rapidly throughout Kent and Sussex and caused severe damage. It was named the 'virulent' or 'progressive' form of the disease, the original strain having been called the 'mild' or 'fluctuating' strain. The two are indistinguishable in laboratory culture, but can be separated from each other by host plant reactions. The disease persists in the soil, the spores of the virulent form surviving there for up to four years. Infection of new plants is *via* the root, and the fungus can be harboured by a number of weed species, which themselves do not show any disease symptoms.

At present there is no chemical agent available for the control of wilt, and eradication of the disease will mean that the infected crop has to be dug up and burned. Land that has carried a wilt-infected crop cannot be replanted except under Government licence. Thus, to combat the disease, hygiene is all-important.

Some hop varieties, such as Bramling Cross, are naturally resistant to wilt, whilst others, such as Wye Target, have had wilt resistance bred into them.

Downy Mildew

Downy mildew (*Pseudoperonospora humuli*) was introduced into Britain in the 1920s, although it had been described from the Far East and the USA earlier in the century. The fungus over-winters in the rootstock and new growth becomes infected. Diseased plants are stunted and spike-like. During summer, black spores are produced (conidiospores) which can directly infect the vegetative parts of other plants. Even strobili can be infected with conidiospores and so yields are severely reduced. Initial treatments involved the use of copper salt solutions, which, in certain circumstances, are still used. Some more recent systemic fungicides are actually curative as well as protective (*e.g.* metalaxyl). Like other downy mildews, high relative humidity is necessary for the proliferation of the fungus and so epidemics will only arise under certain conditions, and then only if no control measures have been taken.

Powdery Mildew

Powdery Mildew (*Sphaerotheca humuli*), sometimes commonly known as 'mould', was first recorded last century and caused severe crop losses in those days. The disease has now largely been contained by fungicides, the first of which was elemental sulfur (used pre-flowering). Sulfur has now mostly been superseded by more effective synthetic compounds. Once established, the disease is difficult to eradicate and, as with many diseases, preventative measures are most effective. The first symptoms of the disease are white leaf-pustules, which give way later in the season to red spots. The white pustules are caused by asexual spores (conidiospores), whilst the red colours are imparted by small fructifications (called cleistothecia) which contain sexually-produced spores. When present on cones, cleistothecia are very difficult to eradicate and severe losses may ensue. Cleistothecia are resistant and are capable of over-wintering; they represent the main route for re-infection of the crop in the following spring.

Red Spider Mite

Red Spider Mite (*Tetranychus urticae*) is only a severe problem in hot, dry climates. The first sign of the presence of the mite manifests itself as minute silver spots on the leaves, which will later turn to red-brown if not treated. Organophosphorus acaricides have, more or less, eradicated the organism, although there is some evidence of resistance developing. The mite has a wide host–plant range which makes control rather more difficult.

Damson Hop Aphid

Damson Hop Aphid (*Phorodon humuli*) has proved to be more of a pest in England than elsewhere, especially before chemical control became possible late last century. Nicotine solutions formed the basis of the earlier sprays, but these have been replaced by organophosphorus compounds. Infection of the hop occurs in May, the aphids having over-wintered as an ovum in the bark of various *Prunus* spp. The hop is the only summer host for the insect. Sprays are normally applied to the leaves of the hop; any aphids becoming lodged in the cones themselves during the flowering period are sheltered and, hence, unaffected by the spray. Untreated plants can become defoliated by mid-summer. Surviving aphids secrete honeydew which can act as a food source for spoilage fungi. Much work is now being carried out on biological control mechanisms.

Viruses

Several viruses are known to use the hop as a host, but only two of them cause significant crop loss in the UK. Symptoms of viral diseases had been described late last century, before the causative agents had been identified.

The two most damaging viruses are the Hop Mosaic Virus (HMV) and the Arabis Mosaic Virus (AMV), both of which are causative agents of nettlehead disease. The main symptom of nettlehead is stunted bine growth and striping and curling-down of the leaves of the plant. Highly susceptible plants can die within a couple of years.

HMV is most serious in the variety Goldings, with most other varieties being symptomless carriers. The virus is sap-borne and transmitted by aphids (normally the damson hop aphid) when they leave their over-wintering hosts. As flight distances for these aphids are short, widespread infection is unlikely.

The viral particle is rod-shaped and some 650 nm long, and infected plants should be grubbed-up and burned.

AMV is rather different inasmuch as the hop is not its only host. It was discovered by Smith and Markham in 1944 who thought it was a glasshouse disease; the virus was actually isolated from rock cress (*Arabis hirsuta*). It affects a wide range of woody and herbaceous plants. In 1966, Bock[25] connected the virus with hop nettlehead symptoms, and also related the disease to what was then called Prunus Necrotic Ringspot Virus (NRSV). The AMV virus has now been linked with split-leaf blotch and bare-bine symptom in certain hop varieties; both of these diseases cause considerable reduction in cone production. The virus, which is an icosahedral particle some 30 nm diameter, is transmitted through the soil by nematodes of the genus *Xiphinema*.

The previously-mentioned NRSV is a problem in continental Europe, especially Germany and the Czech Republic. Hops grown in English climates may carry the virus but there are no outward symptoms of the disease. There at least two Hop Latent Viruses (HLVs), one of which is indigenous to the USA. The only tangible sign of these viruses is a slight lowering of the α-acid content of the host.

REFERENCES

1 J.G.D. Clark and H. Godwin, *Antiquity*, 1962, **36**, 10.
2 R. Ponsonby, *The Brewer*, 1996, **82**, 14.
3 I.R. Ibbotson, *Ferment*, 1998, **11**, 322.
4 R.A. Neve, *Hops*, Chapman & Hall, London, 1991.
5 H.L. Grant, *US Patent*, 4 123 561, 1978.
6 H.L. Grant, 'Proceedings of the 17th European Brewing Convention', Berlin, 1979, p. 441.
7 R.P. Combes, *The Brewer*, 1998, **84**, 29.
8 H. Lermer, *J. Prakt. Chem.*, 1863, **90**, 254.
9 W. Wollmer, *Chem. Ber.*, 1925, **58**, 672.
10 H. Weiland, *Chem. Ber.*, 1925, **58**, 2012.
11 H. Weiland, *Chem. Ber.*, 1925 **58**, 102.
12 C. Lintner and J. Schnell, *Z. Ges. Brauwesen*, 1904, 668.
13 F. Rigby and J. Bethune, *J. Inst. Brew.*, 1957, **63**, 154.
14 M. Verzele, *Bull. Soc. Chim. Belges*, 1958, **67**, 278.
15 M. Verzele and H. De Poorter, *Bull. Soc. Chim. Belges*, 1959, **68**, 85.
16 D. De Keukeleire and M. Verzele, *Tetrahedron*, 1970, **26**, 385.
17 F. Rigby, *J. Chem. Soc.*, 1951, 795.
18 G.A. Howard, J.R.A. Pollock and A.R. Tatchell, *J. Chem. Soc.*, 1955, 174.
19 M. Verzele, in *Brewing Science, Vol. 1*, ed. J.R.A. Pollock, Academic Press, New York, 1979, p. 279.
20 W. Riedl, *Chem. Ber.*, 1952, **85**, 692.
21 W. Riedl, *Annalen*, 1954, **585**, 38.
22 W. Riedl, *Annalen*, 1954, **585**, 209.
23 W. Riedl, *Angew. Chem.*, 1958, **70**, 343.
24 V. Salac, *Wall. Lab. Commun.*, 1955, **18**, 208.
25 K.R. Bock, *Ann. Appl. Biol.*, 1966, **57**, 131.

Chapter 4

Wort Boiling and Cooling

WORT BOILING

The earliest known vessels used for boiling worts were iron cauldrons, which were heated over open fires in mediaeval times. As production volumes increased, heavy, non-malleable cast iron gave way to more flexible copper, which was also more resistant to corrosion and had better conductivity. Thus, the name 'copper' became widely used to describe the wort boiling vessel, a synonym being 'the kettle'. The metallurgical properties of copper enabled elaborate vessels to be constructed, one of the first important developments being the provision of a chimney to carry steam away from the brewing building. Since the 1960s, stainless steel has become more widely used, mainly on the basis of cost and ease of cleaning. In recent years it has been shown that copper actually catalyses some polyphenol oxidation reactions during the boil which lead to colour enhancement.

Coppers were originally direct-fired (*i.e.* over open coal fires housed in furnaces). These were rather inflexible, and to facilitate their use the brewery would normally be equipped with several small coppers, each one being partially filled with wort before the fire was made up (to avoid charring).

Indirect firing, because of its relative efficiency, gradually became the generally acceptable way of wort boiling; the earliest equipment had steam coils situated inside, or steam jackets situated outside, the copper. Coal has now given way to oil or natural gas as a means of boiler heating. Recent advances in the fields of heat transfer and energy economy have led to some very sophisticated wort-boiling systems, some of which incorporate a calandria system whereby wort is circulated over an external heat source. Most modern boiling facilities are designed to minimise heat loss and maximise heat recovery. Nethergate's latest copper is gas-fired, the burning gas being passed through a coiled internal tube (Figure 4.1). Whilst the boil is up, worts are continually being recirculated over the heated coil. Thus, the apparatus is, in effect, a copper-whirlpool (Figure 4.2). Compare this with the unique open copper at Wadworth & Co., Devizes, illustrated in Figure 4.3.

Boiling times in the brewery vary from 45 to 120 minutes, according to the equipment being used, the method of hop usage and the beer type to be produced. Boiling times can be shortened by using slightly enhanced pressures in the copper, although this method is unsuitable for some types of beer. Nowadays, the most widely used source of heat is pressurised steam.

No matter how the copper is designed, the aims of wort boiling are nine-fold:

(1) sterilisation of the wort;
(2) termination of any malt-derived enzyme activity;
(3) concentration of the wort (*i.e.* evaporation of water);
(4) completion of chemical reactions that commenced during mashing, with a resultant lowering of pH;
(5) coagulation of proteins and tannins;
(6) decomposition of unwanted volatile components;
(7) bittering of the wort, due to isomerisation of hop resins;
(8) enhancement of the colour of the wort;
(9) extraction of essential oils and polyphenols from hops.

Some of these reactions occur within a few minutes of the start of the boil (*e.g.* termination of enzyme activity).

(1) The sterilisation aspect of the boil is self-evident; any micro-organisms present in wort will be destroyed at 100 °C. It is, therefore, essential that all brewing equipment post-copper is scrupulously clean in order to maintain sterility.
(2) Most of the enzymes present during mashing are inactivated by the temperature imparted to the mash by the sparge liquor (up to 80 °C). A small percentage will persist, however, and these will be totally inactivated by coagulation during boiling. Any other proteins emanating from the malt will also be destroyed. The boil, in effect, fixes the composition of the wort.
(3) Depending upon the boiling regime, between 5 and 15% of the original wort volume can be lost by evaporation. The steam

Figure 4.1   *View inside the copper at Nethergate Brewery, showing the coiled rings through which burning gas passes*

Figure 4.2   *View of top of copper at Nethergate Brewery. Two of the lids have been removed to show the whirlpool action. Photograph taken just prior to boiling*

Figure 4.3   *The open copper at Wadworth & Co., Devizes*
(Courtesy of Wadworth & Co. Ltd, Devizes, Wilstshire, UK)

emitted will carry some volatile compounds, principally emanating from hop addition.

(4) Boiling completes the process of calcium phosphate [$Ca_3(PO_4)_2$]
precipitation and bicarbonate ion (HCO_3^-) dissociation. Both of
these processes result in a slight pH reduction (normally in the
region of 5.4 to 5.2) owing to the formation of sulfuric acid. As
with mashing, it is the relative proportions of Ca^{2+} and CO_3^{2-}
that actually determine the dynamics of the pH-lowering
reactions; Ca^{2+} is advantageous, CO_3^{2-} is deleterious. Some
brewers actually add sulfuric acid to the copper to obtain a
specified pH. The reduction of pH during boiling encourages the
precipitation of proteins.

The completion of the precipitation of calcium oxalate also occurs during coppering.

(5) Coagulation of proteins is important during boiling because any that survive will end up in the final beer. Some are required to impart 'body', but otherwise they can cause problems. Heat causes the proteins to become permanently uncoiled, whence they react with polyphenols and become the sedimented out as hot trub particles. The nearer the pH of the wort to the isoelectric point of the protein concerned, the more rapid the sedimentation will be. Ca^{2+} and a vigorous boil also encourage precipitation. Because of the nature of the reactions involved, protein coagulation is one of the slower processes to occur during boiling, and, for this reason, most breweries add 'copper finings' to the boil, usually towards the end. The earliest forms of such fining material were dried samples of certain species of red algae (Rhodophyta), generally sold under the name of 'Irish Moss'. The main species used were *Chondrus crispus* and *Gigartina stellata*, both common lower littoral seaweeds. Specially prepared extracts are now used, usually in pellet form, whereby the active phycocolloid is dispersed in a stabiliser. The all-important ingredients in these algal extracts are the negatively-charged, sulfated polysaccharides, carragheenan being the principal one.

Alginic acid and its sodium salt are also now used for protein flocculation in the copper. The acid is found in most of the larger species of brown algae (Phaeophyta), most commercial extraction coming from the genera *Laminaria*, *Fucus* and *Ascophyllum*.

The elementary hot trub particles have a diameter in the range 0.5–1.5 μm, but after the turbulence of the boil they flocculate to produce much larger masses, which can be up to 1 cm in diameter. The mass of hot trub is referred to as the 'hot break'. After boiling, the trub and hop debris must be separated from the bulk of the hot, hopped wort (which will be cooled in the next stage of the brewing process). The earliest means of separation involved sieving and straining and were very laborious.

As breweries increased in size and production the hop back was developed, which essentially acted as a sieve. Often situated underneath the copper, the hop back in its simplest form resembles an infusion mash tun, being an open vessel with raised perforated plates at the bottom. The contents of the copper would be 'cast' into the hop back, whereby the leaf hop debris and the hot break material would be held on the plates whilst the hot wort would pass through to the chiller. Most hop backs have a sparging apparatus so that all the wort can be removed from the debris. One of the original ways of late-hopping involved adding fresh aroma hops to the hop back whilst the copper was being cast. Some breweries recirculate copper castings over the hop back bed to ensure maximum trub removal.

With the advent of hop pellets and hop powders the volume of debris in the copper after boiling is far lower than it would be if leaf hops are used. Moreover, the debris is more finely divided and almost indistiguishable, size-wise, from hot break material, so separation in a hop back – or any other type of sieve – would be impractical. There are advantages, however, especially in the field of brewery engineering. Firstly, pumps can be used to move boiled wort prior to the removal of hop debris and hot trub. This facility led to the development of the whirlpool separator, whereby worts are pumped tangentially into a vertical tank and left to circulate. This results in the solid material accumulating in a mass at the centre of the bottom of the vessel; clear worts can then be drawn off from above the deposit. Deposition can be enhanced if the bottom of the separator vessel is conical.

Other methods of removing fine debris from boiled wort include centrifugation and filtration; these methods are mainly applicable to large breweries.

Spent hops and hot break material have reasonable nutritive value and may be sold for animal feed, either solely or with spent grains from the mash tun.

(6) During boiling, many volatile substances are lost with the main flow of steam; some of these are undesirable in terms of their flavour and aroma contributions. Unfortunately, some desirable volatiles are lost as well but the brewer can partially compensate for this by making a late hop addition to the copper.

(7) When leaf hops or pellets are added to a boiling copper the first stage in the formation of bitterness is the extraction of resins from the lupulin glands. The resins are not very water-soluble, but they slowly dissolve in hot wort before the isomerisation reactions take place. Extraction of resins into wort is an inefficient process and rarely exceeds 50%, even in the most modern breweries.

The degree of extraction is usually expressed in terms of

percentage hop utilisation, and this can be determined for worts and for beer using equation 4.1.

$$\% \text{ Hop utilisation} = \frac{\text{concn. of iso-}\alpha\text{-acids in wort/beer}}{\text{concn. of }\alpha\text{-acids added to wort}} \times 100 \quad (4.1)$$

Concentration is usually expressed in terms of mg litre^{-1}. Hop utilisation rates in finished beer may be as low as 20% and will rarely be more than 40%. Several factors affect hop utilisations in the copper, the most important being the rate of hopping and the vigour and length of the boil. Evidence suggests that most isomerisation occurs within 60 minutes at 100 °C. Hop pellets and shredded cones give better utilisation rates than intact cones, indicating that it is the dispersal of resins during the boil that gives greater bitterness. Dispersal is encouraged by boiling vigorously. High hop rates give a lower percentage utilisation than lower hop rates, whilst high gravity worts produce lower hop utilisation rates than worts of lower gravity.

As stated previously, during the boil the α- and β-acids are isomerised to form their iso-derivatives. The α-acids seem to be isomerised more rapidly than their β-counterparts and are certainly the major contributors to bitterness. Much *in vitro* research has been carried out on these isomerisations, particularly that of humulone where it has been shown that, at constant pH, the reaction follows first order kinetics. This is evidently not what happens in the copper because pH is not constant (acidity is increased slightly during boiling). The mechanism for the isomerisation of humulone is given in Figure 4.4 and, in chemical terms, we are observing an acyloin, or benzilic acid, rearrangement. The humulone–isohumulone rearrangement is reversible, but the two isohumulone isomers are not interconvertible, except *via* humulone itself. The absolute configurations of *cis*- and *trans*-isohumulone were established by De Keukeleire and Verzele in 1971.[1] Conversion of the other α-acids appears to follow the same principles.

Apart from the reactions depicted in Figure 4.4, many other compounds are produced, in fractional amounts, during wort boiling. These include the humulinic acids (1) and the humulinones (2).

The β-acids are converted to their iso-compounds in a similar fashion, although the rearrangements take place more slowly and to a lesser extent than for α-acids.

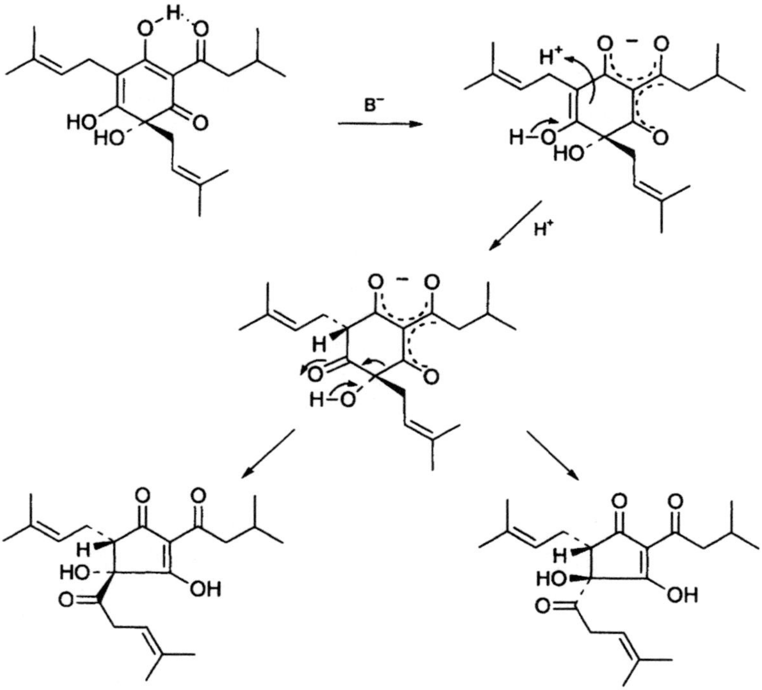

cis-Isohumulone

trans-Isohumulone

Figure 4.4 *Isomerisation mechanism of humulone* (Courtesy of the Institute of Brewing)

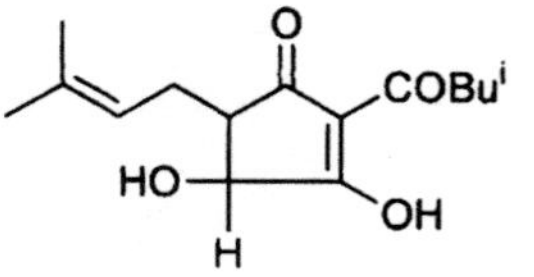

Humulinic acid

(1)

A hydrolysis product of humulone.

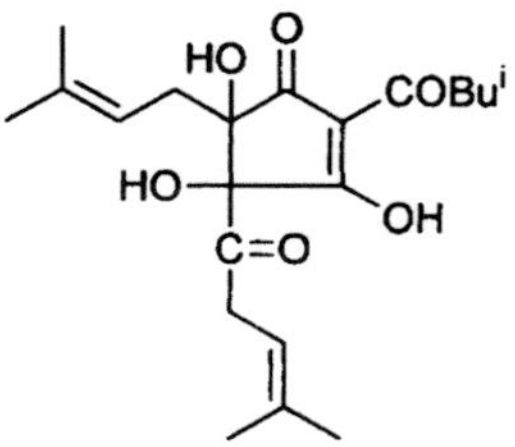

Humulinone

(2)

Can be formed by oxidation of humulone with organic peroxides. It is intensely bitter.

For a comprehensive treatment of hop constituents, and their fate during wort boiling, the reader is referred to the extensive works of Verzele,[2,3] Hough *et al.*[4] and Stevens.[5]

(8 and 9) The colour of wort intensifies during the boil, and prolonged boiling, especially under pressure, will produce marked darkening. Elevated pH and oxygen also promote colouring. Two main types of colour enhancement occur: (a) oxidation of polyphenols; and (b) interaction of carbohydrates and nitrogenous compounds.

(a) Simple and polymerised phenols are extracted into boiling wort; these compounds emanate from both malt and hops. As mentioned previously, a proportion of these compounds will react with proteins to form the hot break. Those that do not, undergo a variety of interactions, where some of which produce coloured compounds. Oxidation of phenols and polyphenols certainly occurs, with the subsequent formation of quinones and their derivatives (which themselves are oxidising agents).

A wide range of phenolics have been reported from boiled wort, and these compounds were originally classified by brewing chemists as follows:

(i) Phenolic acids. These are mostly derivatives of hydroxybenzoic acid and hydroxycinnamic acid. Gallic acid (3), ferulic acid, vanillic acid, *p*-coumaric acid and chlorogenic acid are probably the most important in this class, the last three being present in malt as well as hops. All of the phenolic acids are only present in trace amounts and are now regarded as being insignificant in terms of beer flavour and stability. Only vanillic acid, *p*-coumaric acid and ferulic acid seem to be universally found in finished beer.[6]

Gallic acid
(3)

(ii) Flavanols, of which kaempferol (4) and quercetin are most widely distributed. Flavanols are hop-derived and form a wide variety of glycosidic derivatives. According to

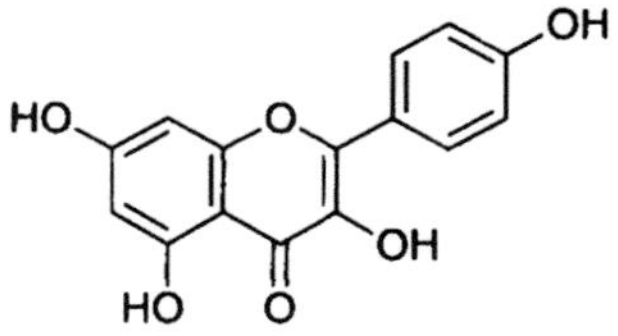

Kaempferol
(4)

McMurrough and Delcour,[7] there is little evidence that these compounds actually have any effect on the flavour, colour or stability of beer.

(iii) Anthocyanogens and related compounds, which on hydrolysis yield coloured compounds collectively known as anthocyanidins. There are two groups of anthocyanogens according to the number of flavanoid units involved. Those derivatives with one unit are called leucoanthocyanidins, whilst those with two or more are designated proanthocyanidins. In terms of organic chemistry, these are all flavanoid polyphenols. Such compounds are all related to the polyhydroxyflavan catechin (5), which is widely distributed in plants and may be considered to be a monomeric flavanol.

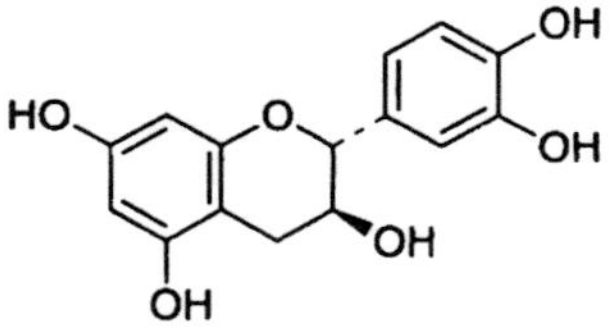

Catechin
(5)

The oligometric flavanoids consist of ordered chains of polyhydroxyflavan-3,4-diol monomers, with either (+)-catechin, or its close relative (−)-epicatechin, as terminal units. In brewery terms, the most important of these small polymers are the proanthocyanidins, of which the dimeric procyanidin B3 and prodelphinidin B3 are the most significant, being found in barley, malt and hops. These simpler flavanols are readily extracted from raw materials during the brewing process, but only a small proportion remains unaltered and ends up in the final beer (where it causes instability problems).

(b) Interactions of carbohydrates and nitrogenous compounds. During boiling, wort colour intensifies as a result of a series of reactions between reducing sugars and primary amines. These 'browning' reactions are non-enzymatic and involve a number of intermediate compounds, some of which are unstable. The pigments produced are polymeric, although their precise structures have not been elucidated. They resemble the caramels in many respects and are chemically referred to as melanoidins. Melanoidins are water-soluble, but insoluble in many organic solvents. The reactions leading to their formation are of the Maillard type and involve Amadori rearrangements and Strecker degradations.

During a normal wort-boiling period only about 10% of the amino acids and sugars undergo Maillard reactions, but this is sufficient to impart flavour and colour characteristics to the beer. If boiling were prolonged the logical conclusion would be the production of caramel itself. Commercially-produced caramel can be used to impart colour artificially to beers, and this was a widely-used practice until a few years ago. A wide range of melanoidins are produced, depending upon the amino group and sugar involved. Maillard reactions also occur during the kilning of malt.

The phenomenon of non-enzymatic browning should not be confused with the other, biological, browning reaction that results in the formation of melanins. This is enzymatically mediated and is brought about by the action of polyphenol oxidase on an appropriate substrate, such as tyrosine. Oxygen is also required in this reaction, whereas non-enzymatic browning is not necessarily oxygen-dependent. For a comprehensive review of non-enzymic browning, see Reynolds.[8,9] Many other compounds result from browning reactions, and although their contribution to colour might be minimal, some of them such as the pyrroles, thiazoles and furfuryl compounds do enhance beer flavour.

WORT COOLING

Hopped wort, which has been separated from hop debris and hot trub, is now chilled before being introduced into fermentation vessels and pitched (mixed) with yeast. The earliest commercial breweries used shallow open trays (Figure 4.5), called coolships (these were

Figure 4.5   *Boiling hopped wort being cast into the coolship at Elgood & Sons, Wisbech*
(Courtesy of Elgood & Sons Ltd, Wisbech, Cambs., UK)

usually constructed of copper), through which hot wort was slowly passed. The cooling effect was solely dependent upon ambient temperature and there were obvious inherent problems, especially over the summer months. During cooling, the wort becomes oxygenated and further trub settles out – the cold break. Cold break material is very insoluble, consisting of proteins and polyphenols, and is deposited in the coolship. The time required to bring about cooling and the potential for microbial contamination with this apparatus is obvious and they were gradually replaced by other cooling systems. One such variant was an apparatus whereby hot wort was trickled down vertical metal plates or tubes which were chilled with cold water (usually spring or river water). Such chillers could effect wort cooling and oxygenation, but the cold break material could not be removed and would pass into the fermentation vessel if not removed by filtration or centrifugation. Again, because the wort was passing over the chilled surface and in contact with air, there was a likelihood of infection.

Nowadays, most breweries (including Nethergate) use plate heat-exchangers (or paraflow chillers – see Figure 4.6) whereby a series

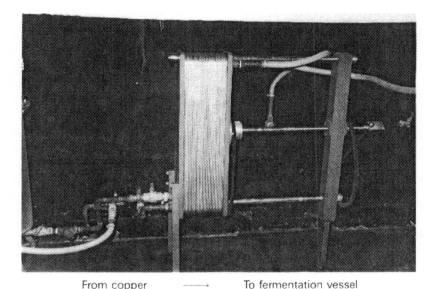

From copper ⟶ To fermentation vessel

Figure 4.6 *The wort cooler (paraflow) at Nethergate Brewery*

of vertical, indented, stainless steel plates bearing rubber gaskets are compressed together in a frame. Each plate has four circular holes at its corners and, when they are held tightly together, the grooved indentations and the circular holes form a series of channels through which hot wort can pass in one direction, and the coolant (normally chilled water) in the other. The chillers, which are very expensive, are designed to facilitate a turbulent flow over as large a surface area as possible Such systems are closed and very efficient, but they do not oxygenate wort or remove cold trub. An economic advantage of the paraflow is that, through cooling the wort, large volumes of hot water are produced which can be retrieved and subsequently used for brewing or cleaning down. To improve the efficiency of heat-exchangers, ethylene glycol or alcohol may be used as a refrigerant.

The final temperature of the cooled wort will vary according to the type of beer being produced, *i.e.* the fermentation temperature required. For lager-style beers, worts are passed to the fermenter at 10–15 °C; for ales, temperatures of 16–20 °C are necessary.

Passage of some cold trub to the fermenter cannot be avoided, but this does not normally cause problems during ale fermentations; in fact, there is evidence to suggest that it can be beneficial. Lager brewers, however, normally remove the bulk of the cold break material by centrifugation, filtration through Kieselguhr, or by a flotation method whereby cooled wort is kept in a vessel through which air is pumped. Air bubbles and cold trub float to the top of the vessel where the latter can be skimmed off.

As we shall see, oxygen is essential for the early stages of yeast growth; the precise requirement for oxygen will vary with yeast strain. If the cooling process has not permitted sufficient oxygen to enter the cooled wort, then artificial oxygenation must be carried out. This can be achieved by injecting oxygen at either the hot wort end, or the cold wort end of the heat-exchanger. Injection at the hot end results in less physical solution of oxygen and a darker wort colour than if injection is post-paraflow. To avoid any undue oxidative changes in the wort, it is common practice to inject air on the hot side of the paraflow and pure oxygen on the cold side.

REFERENCES

1 D. De Keukeleire and M. Verzele, *Tetrahedron*, 1971, **27**, 4939.
2 M. Verzele, in *Brewing Science, Vol. 1*, ed. J.R.A. Pollock, Academic Press, New York, 1979, p. 279.
3 M. Verzele, *J. Inst. Brew.*, 1986, **92**, 32.
4 J.S. Hough, D.E. Briggs, R. Stevens and T.W. Young, *Malting and Brewing Science, Vol. 2, Hopped Wort and Beer*, Chapman & Hall, London, 2nd edn., 1982.
5 R. Stevens, in *An Introduction to Brewing Science and Technology*, ed. R. Stevens, Institute of Brewing, London, series II, vol. 1, 1987, p. 23.
6 I. McMurrough, G.P. Roche and K.G. Cleary, *J. Inst. Brew.* 1984, **90**, 181.
7 I. McMurrough and J.A. Delcour, *Ferment*, 1994, **7**, 175.
8 T.M. Reynolds, *Adv. Food Res.*, 1963, **12**, 1.
9 T.M. Reynolds, *Adv. Food Res.*, 1965, **14**, 167.

Chapter 5

Fermentation

THE YEAST

The most widely used yeasts in the brewing industry are members of the fungal genus *Saccharomyces*, of which there are over 30 known species. In the broadest sense, fungi (Mycota) fall into the kingdom Protista, as defined by Haeckel. Because of the paucity of undisputed evidence, the evolutionary origins of fungi remain a matter of conjecture. Whilst many authorities regard them as protists, there are some who assign fungi to their own separate kingdom of organisms.

They are certainly eukaryotic organisms, possessing double unit membrane-bound organelles (such as nuclei and mitochondria). The true fungi (Eumycota) possess a rigid cell wall, which allies them to the plant kingdom; but, they are non-photosynthetic and so are forced into a heterotrophic mode of existence.

Within the Eumycota, yeasts fall within the sub-division Ascomycotina (ascomycetes); the largest single group of fungi. They are characterised by the production of sexual (or perfect) spores, called ascospores, which are contained within a sac, or ascus. The majority of ascomycetes produce a spore-bearing structure, or fructification (ascocarp), to support and protect their ascospores; these are placed in the Euascomycetes (true ascomycetes). Those that do not produce ascocarps are classified as Hemiascomycetes; the yeasts and their allies belong to this latter group. Most Hemiascomycetes are unicellular organisms, although some do produce filamentous growths (hyphae).

Saccharomyces spp. are located within the family Saccharomycetaceae and are distinguished from each other on morphological and physiological (growth) characteristics. True yeasts reproduce vegetatively by budding (Figure 5.1), under certain conditions. When a bud

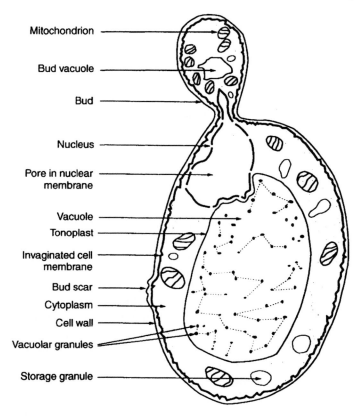

Mitochondrion

Bud vacuole

Bud

Nucleus

Pore in nuclear
membrane

Vacuole
Tonoplast

Invaginated cell
membrane

Bud scar
Cytoplasm
Cell wall

Vacuolar granules

Storage granule

Figure 5.1   Saccharomyces cerevisiae, *diagrammatic representation of a*
section of a budding yeast cell as seen under an electron microscope
(Taken from *Introduction to Fungi*, J. Webster. © (1970)
Cambridge University Press)

is extruded a scar is left on the wall of the mother cell. In theory, there
is enough cell wall surface area on a mature cell for about 100 buds; in
practice, no more than 50 scars have ever been observed.

Some 75–80% of the yeast cell consists of water, whilst on a dry
weight basis the actively growing cell comprises:

—*ca.* 40% protein, mostly in the form of enzymes (of which over
1000 have been identified) but also located in the cell wall and
bound to membranes;
—*ca.* 34% polysaccharide, mainly in the cell wall and as food
reserve material;

—7% minerals (called the ash content);
—5% phospholipids;
—3% triglycerides;
—0.5% DNA, vitamins and fibre.

The cytosol (or cytoplasm) is bounded by a cell envelope which consists of cell membrane, periplasmic space and cell wall. The cell wall can be 150–300 nm thick, according to strain and growth conditions, and is metabolically active, containing enzymes capable of permitting macromolecular transfer into the cell. The membrane and periplasmic space are both 10 nm thick. Up to 85% of the cell wall dry weight is attributable to two structural polysaccharides which are present in roughly equal amounts; these are:

(1) β-Glucans; glucose polymers which comprise the inner layers of the wall and are responsible for cell shape and wall rigidity.
(2) Mannoproteins (α-mannans); mannose polymers linked covalently to peptide chains, constituting the outer layer of the wall and responsible for porosity and environmental reception.

The cell surface of the top-fermenting (ale) yeast is covered in small microfibrillar protrusions (thought to be mannoprotein) which confer a roughness that enables the cells to rise to the surface during fermentation. In between the mannoprotein outer layer and the glucan-dominated inner layer there is a series of intermediate layers composed of both entities.

In addition to the two main polymers, chitin (a polymer of *N*-acetylglucosamine) is present in small amounts (*ca.* 1% dry weight). Chitin is a major component of many fungal cell walls, but in the yeast wall it appears to be concentrated in the bud scars. Small quantities of protein, lipid and inorganic phosphates are also incorporated into the wall matrix.

In addition to a rough surface in some forms, the yeast cell wall surface also possesses a net negative charge and exhibits hydrophobicity; all three characters are of importance to the brewer. The negative charge is attributable to phosphate chains located in the mannoprotein outer wall and can be demonstrated by alcian blue staining. The extent of the charge can be measured by ion-exchange chromatographic or electrophoretic (zeta-potential) methods. The charge is particularly important during the fining of cask beer, when the positively-charged sites on the collagen molecules in isinglass (a material used for fining) attract individual yeast cells and promote sedimentation. The strength of the charge varies with environmental conditions (such as starvation) and during fermentation there is a reduction in the charge at the onset of flocculation.

Hydrophobicity is conferred by lipids in the outer wall matrix and phosphorylated groups in the mannoprotein complex. It may be determined by means of solvent partition assay and affinity chromtography. Again, the degree of the incapacity of the yeast cell wall to interact with water will be determined by environmental circumstances. Hydrophobicity increases immediately prior to flocculation and is thought to be very important in this process.

The cell membrane appears invaginated under the electron microscope and in frozen-etched surface view these invaginations can be seen as elongated folds, some 4 μm long. Also visible are clusters of particles about 180 Å diameter; these are arranged hexagonally. The membrane is composed of lipids and phospholipids with associated proteins and sterols (steroid alcohols).

The most conspicuous organelle within the yeast cell is the vacuole (Figure 5.1). This is bounded by a single unit membrane, the tonoplast, which is slightly thinner than the plasmalemma. The vacuolar sap is rich in volutin (polymetaphosphate) granules during periods of metabolic inactivity. These dense granules disappear when active growth recommences. The vacuole also acts as a store for lytic enzymes involved in the recycling of cell macromolecules.

The second largest organelle is the nucleus, which is typically double unit membrane-bound, with nuclear pores. The nucleus is some 1.5–2.0 μm in diameter, whilst the pores are of the order of 0.1 μm diameter. Within the nucleus there is a darker, denser area called the nucleolus. It is not delimited from the rest of the nuclear sap by a membrane. The nucleolus is rich in RNA and protein, and high resolution electron microscopy indicates that there are two distinct regions; one particulate, one fibrillar. The nucleolus is the site of synthesis and organisation of cytoplasmic (80S) ribosomes.

Yeast chromosomes are very small and difficult to discern. There are still conflicting reports on the chromosome number of the diploid cell. Early communications indicated that it was eight, but this is not in accordance with information obtained from genetic research, where it has been shown the nucleus of the haploid cell possesses at least 17 linkage groups.

Connecting the plasmalemma with the nuclear membrane is a diffuse, double unit membranous system; the endoplasmic reticulum (ER). The form of the membranes can vary, morphologically, from tubular, to vesicular, through to flattened sacs (cisternae), and the actual form and amount of ER is dependent upon the condition and age of the cell. Some ER has ribosomes attached to the membranes giving it a rough appearance. In effect, the ER divides the cytoplasm of the cell (the content of the cell apart from the nucleus) into two main regions; that which is enclosed within the membrane system and that which is exterior (the cytoplasmic matrix).

The cytoplasm also contains Golgi bodies (dictyosomes), which are, again, double unit membrane-bound. They are implicated in cell wall growth and at certain stages of the cell cycle they are difficult to distinguish from the ER.

Mitochondia are liberally dispersed in the cytoplasm and, as in other eukaryotes, they contain their own (70S) ribosomes and are the sites of adenosine triphosphate (ATP) production.

Other cellular organelles found in *Saccharomyces* spp. are lipid granules (sphaerosomes) and lysosomes, which contain hydrolytic enzymes.

Reproduction in the yeast is either asexual, by vegetative budding, or sexual; a process involving mating and the ultimate production of sexual spores (ascospores).

Budding is a mechanism unique to *Saccharomyces* spp. It is initiated by a weakening of the existing cell wall in a localised area; this is brought about by lytic enzymes. Vesicles from the ER congregate in this area, which becomes one of high metabolic activity. Internal pressures force the cytoplasm through the weakened wall and this 'bulge' immediately becomes surrounded by newly synthesised cell wall material. The nucleus migrates to the bud-forming site and, when a complement of cell organelles have passed into the embryonic bud, a septum of cell wall material separates the bud from the mother cell. This septum actually constricts and pinches off part of the nucleus, which, in effect, produces a 'bud' of its own. The nuclear membrane persists at all times during this process. Chromosomal events are unclear, but mitosis will have occurred during the bud-forming process.

As cells grow and divide (in this case by budding), the period between the end of one division and the end of the next is called the cell or replication cycle (Figure 5.2). There are two phases to the cell cycle: the growth phase (where the nucleus is in interphase) and the division phase. The growth phase, itself, is divisible into three distinct periods, each one having its own characteristic molecular events. The stage immediately following the last cell division is called G1 and the cell in this phase is 'young', having just undergone mitosis. In terms of the yeast cell, G1 represents the part of the cycle immediately

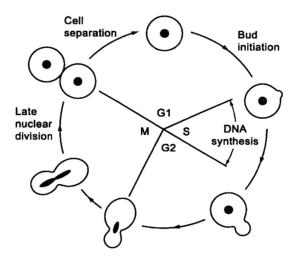

Figure 5.2   *Cell cycle of* Saccharomyces cerevisiae
(Courtesy of Dr D.E. Briggs, Dr R. Stevens and Dr T.W. Young)

following the excision of a bud up until the initiation and emergence of a subsequent bud. The latter could be formed on the mother cell or on the bud itself. The second period of the growth phase is the S stage. Here, DNA and other chromosomal material is replicated. The morphological manifestation of the S stage in yeast is the increase in size of the bud. As soon as DNA synthesis has been completed, the cell, by definition, enters the the G2 phase. In *Saccharomyces* spp. this lasts until the nucleus has migrated into the constriction between mother cell and bud. G2 is immediately followed by the M phase (meiosis), whereby the chromosomes are duplicated. The M phase progresses through the normal mitotic events (*i.e.* prophase, metaphase, anaphase and telophase) until the bud actually separates, whence G1 ensues again.

Brewing yeasts can exist, vegetatively, with a single set of chromosomes (haplophasic) or a double set of chromosomes (diplophasic), as illustrated in Figure 5.3. Both phases are capable of proliferation by budding. Under appropriate conditions haploid cells will mate, their cytoplasms will fuse (plasmogamy) and this will be followed immediately by fusion of nuclei (karyogamy). By definition, the result of karyogamy is the production of the zygote. Gradually, the zygote will develop into a diploid vegetative cell,

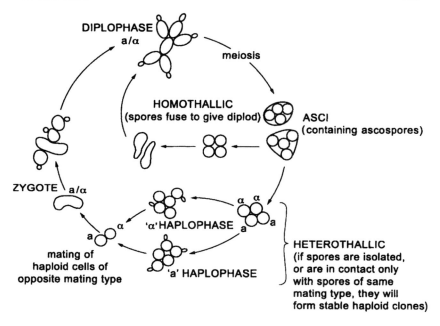

Figure 5.3   *Life cycle of* Saccharomyces cerevisiae *(a and α refer to the genes controlling mating response)*
(Courtesy of Dr D.E. Briggs, Dr R. Stevens and Dr T.W. Young)

which, under most conditions of growth, will be morphologically similar to the haploid vegetative cell. Many strains of *Saccharomyces cerevisiae* are heterothallic, there being two mating types involved. Mating type is controlled by a single gene which exists in two allelic states, a and α. These heterothallic types are of vital importance in the field of yeast genetics. The opposite condition, in terms of the mating of haploid cells, is homothallism, whereby only one strain is necessary for zygote production. Homothallism is determined by the dominant HO allele.

In laboratory culture, diploid cells will proliferate by budding for as long as there are sufficient nutrients to support the process in the growth medium. When nutrients become limiting the cell will resort to the production of ascospores. Artificially, this can be induced by transferring diploid cells from a rich growth medium to an impoverished one. Ascospores are haploid and thus the diploid cell has undergone reduction division (meiosis) during their formation. The course of meiosis in *Saccharomyces cerevisiae* follows the normal pattern and involves two nuclear divisions: the first involving the replication, breaking and re-joining of chromosomes, ultimately resulting in the formation of two diploid cells; the second resulting in the production of four haploid cells. These haploid cells produce a slightly thickened cell wall and, by definition, become ascospores. The original mother cell that produced them being the ascus. In some species of *Saccharomyces* each of the four haploid cells initially produced *via* meiosis can undergo a subsequent mitotic division, which results in the production of eight ascospores per ascus. Ascospores are smaller and more rounded than normal vegetative cells.

There are two fates for ascospores:

(1) they are liberated from the ascus and develop into haploid vegetative cells, which will proliferate by budding;
(2) they fuse in pairs within the ascus and develop into diploid vegetative cells, which will also proliferate by budding.

The fact that mitosis occurs in the diplophase of yeasts differentiates them from all other eukaryotes, where, normally, mitosis is confined to the haplophase.

Brewing strains of yeast (both ale and lager) have been shown to have rather different characters to strains grown *in vitro*:

—they are usually polyploid (aneuploid);
—they have low sporulation efficiencies;
—when produced, ascospores have low viability;
—ascospores have low mating efficiency.

THE NUTRITIONAL REQUIREMENTS OF YEAST

The overall growth requirements of a brewing yeast are:

(1) a carbon energy source, known as fermentable sugars;
(2) a nitrogen source;
(3) growth factors – vitamins;
(4) inorganic ions (other elements);
(5) oxygen (especially during the early stages of fermentation);
(6) water.

Carbon Metabolism

The major sugars in an all-malt wort are expected to exhibit the following profile, in terms of percentage composition:

Sugar	% composition
Maltose	50–60
Dextrins	20–30
Maltotriose	15–20
Glucose	10–15
Fructose	1–2
Sucrose	1–2

Some are taken up passively by the cell in an intact form (*e.g.* glucose and fructose), some are hydrolysed outside the cell and the breakdown products are absorbed (sucrose), whilst others are actively transported across the cell membrane and hydrolysed in the cytosol of the cell (maltose and maltotriose). Dextrins, comprising maltotetraose and larger starch breakdown products, are not metabolised. Any minor amounts of pentose sugars are also left unfermented.

The general pattern of disappearance of fermentable sugars from wort during fermentation is sucrose–glucose–fructose–maltose–maltotriose, although there are differences between yeast strains. Details of carbohydrate utilisation are given under the section on fermentation on pages 115–117.

Nitrogen Metabolism

Brewer's yeast cannot assimilate elemental nitrogen or nitrate ions. Some strains can utilise ammonium ions but the bulk of the nitrogen required for the synthesis of essential cell constituents comes from the amino acids and the di- and tri-peptides in the wort. These moieties will have originated from malt itself.

In order to achieve a good, rapid, fermentation the α-amino nitrogen content of wort should not be less than 100 mg l^{-1}, preferably in the region of 150–200 mg l^{-1}. There should also be a balance of amino acids for the yeast to utilise.

As is the case for carbohydrate utilisation by yeast, amino acids are taken up and utilised sequentially, according to the presence, or otherwise, of the appropriate transport enzymes in the membrane. A typical all-malt wort contains some 19 amino acids and these may be placed into four groups according to their rate of uptake by yeast:

(1) fast absorption: glutamic acid, aspartic acid, asparagine, glutamine, serine, threonine, lysine, arginine;

(2) intermediate absorption: valine, methionine, leucine, isoleucine, histidine;

(3) slow absorption: glycine, phenylalanine, tyrosine, tryptophan, alanine;

(4) little or no absorption: proline (the most abundant of the wort amino acids).

The cell itself produces amino acids by transamination and/or by synthesis from various acids in the keto acid pool. In some yeast strains certain amino acids seem always to be synthesised within the cell, even though they are present in plentiful quantity in wort; arginine, histidine and lysine appear to fall into this category.

Proline necessary for yeast metabolism is synthesised *de novo* from glutamate.

Protein synthesis in yeast is both mitochondrial and cytoplasmic.

Vitamins

Wort provides a rich source of vitamins and, although yeasts vary greatly in their vitamin requirements for growth, there is normally an adequate supply of both variety and quantity emanating from the mash tun.

Wort should contain biotin, thiamin (B1), nicotinic acid, riboflavin, calcium pantothenate, inositol and pyridoxine, pyridoxal and pyridoxamine.

Except for inositol, which is implicated in membrane synthesis (phospholipids), *i.e.* a structural role, all vitamins serve a catalytic function as part of a coenzyme in metabolism (non-functional). Most brewing strains have an absolute requirement for biotin, and many require pantothenate and inositol.

Inorganic Ions (Other Elements)

A number of inorganic ions are required by yeast during fermentation, some performing a structural role and some being required for enzymic purposes.

Sulfur

Elemental sulfur, in colloidal form, can be utilised by yeast under certain circumstances, but this rarely occurs under the conditions of brewery fermentations. Sulfate ions, a universal constituent of wort, can also be utilised but this is on a small scale whilst there are sulfur-containing amino acids in the wort. The uptake of sulfate ions is *via* an active process and most of it will occur early on during fermentation – before the sulfur-containing amino acids have been taken up in sufficient concentration. Methionine is the most easily utilisable of the wort organic sulfur compounds but even this is absorbed slowly during the initial hours of fermentation. Sulfur-containing proteins and peptides are also present in wort, as are the vitamins thiamine and biotin. For fermentation to proceed satisfactorily it is essential that 40–50 mg sulfur per litre of wort be present in organically-bound form.

Apart from the structural proteins and their constituent amino acids, the most important sulfur-containing compounds within the yeast cell are coenzyme A, lipoic acid, thiamine pyrophosphate (TPP) and glutathione. The latter is a tripeptide food reserve (glutamic acid–cysteine–glycine) and represents some 20% of cell sulfur. Its formation is enhanced by excess sulfate ions in the wort, and under certain growth conditions can contribute up to 1% of the cell dry weight.

Of the metabolic by-products emanating from sulfur utilisation during fermentation, hydrogen sulfide (H_2S) is the most significant and is generated in detectable quantities during vigorous phases of yeast growth. Most appears to be produced from the reduction of sulfate ions (SO_4^{2-}), the prevalent acidity permitting its release. Some H_2S, however, does arise from the metabolism of organosulfur compounds in the wort. Most of the H_2S produced is chemically bound, and that fraction that does survive in gaseous form will normally be lost with the bulk of the evolution of CO_2. Thus, H_2S concentration in beers rarely exceeds 1 ppb, which is below its flavour threshold (30 ppb). Easily detectable (smell!) amounts of H_2S in a beer are a sign of brewery infection with bacteria such as *Zymomonas mobilis*.

Sulfur dioxide (SO_2) is present in beer, usually in bound form and always at levels below its flavour threshold [20 000 ppb (20 mg l^{-1})].

The major organosulfur by-products of yeast metabolism are the mercaptans (thio-alcohols) which, with their less abundant derivatives, the thio-aldehydes and thio-ketones, do contribute to the overall beer flavour.

Most volatile sulfur compounds found in beer, as we have seen, originate from hop volatiles and are not a direct result of fermentation.

Metallic Elements

Several metallic ions play an important part in the brewing process, particularly during fermentation. These can be placed into three classes:

(1) *Macroelements.* Required in concentrations between 0.1 and 1.0 mM, this class includes K^+, Mg^{2+}, Ca^{2+}, Zn^{2+}, Fe^{2+}, Fe^{3+}, Mn^{2+}.

(2) *Microelements.* These are required in concentrations between 0.1 and 100 mM, and include such elements as Co^{2+}, Cd^{2+}, Cr^{3+}, Cr^{6+}, Cu^{2+}, Mo^{2+}, Ni^{2+}, V^{2+}.

(3) *Inhibitors.* At concentrations above 10–100 mM, included here would be Ag^+, As^{3+}, Hg^+, Li^+, Os^{2+}, Pd^{2+}, Se^{4+}, Te^{4+}.

During fermentation, the concentrations of Mg^{2+}, Zn^{2+}, K^+ and Co^{2+} are of most significance. The optimum concentration of any one element is partly dependent upon the concentration of other ions; a situation which can cause complications. This is particularly true for the relationship between zinc and manganese. Calcium and magnesium ratios are also known to be critical during fermentation. Most cationic elements are present in wort in several forms, only some of which are bioavailable. Bioavailability of these elements depends upon wort pH, availability of anions, and presence (or otherwise) of chelating agents.

Magnesium. The requirement for this element is small and amply provided for by malt and liquor itself. The minimum requirement for most brewing strains is around 42 ppm (1.7 mM), whilst in concentrations above 25 000 ppm (1 M) it is inhibitory to growth. Yeast actively takes up the element *via* a high affinity specific transport system and it is the most abundant intracellular divalent cation in the yeast cell. Magnesium exerts most of its effect intracellularly and is required for many enzymes, including all kinases, phosphatases and synthetases. It regulates pyruvate metabolism, since the glycolytic enzymes hexokinase, phosphofructokinase, endolase and pyruvate kinase are all Mg-activated. It also serves several other functions: (a) it plays a role in alcohol tolerance; (b) has a protective effect when the cell is under conditions of stress (particularly temperature and osmotic stress); (c) it is involved in the membrane stabilisation of nucleic acids, ribosomes, lipids and polysaccharides; (d) it stimulates fermentation of high-gravity worts; (e) it is involved in ribosome structure; (f) it has a neutralising effect on the electrostatic forces present in polyphosphates, DNA, RNA and proteins; (g) it plays a key role in cell growth and division.

According to Rees and Stewart,[1] 'yeast cells have a high demand for magnesium, which cannot be effectively met by other ions; and that magnesium is fundamental to the metabolic and physiological functions of the cell.'

Calcium. An important element in yeast membrane structure and function. The minimum requirement for most strains is in the region of 10–20 ppm (0.25–0.5 mM), whilst above 1000 ppm (25 mM) it acts as an inhibitor. Its presence can stimulate yeast growth, even though it is not strictly a growth requirement. Most of the influences of calcium are extracellular as far as the yeast is concerned and primarily involve the maintenance of membrane and cell wall integrity. Cytoplasmic levels of Ca^{2+} are maintained at a very low residual level [4–8 ppm (100–200 nM)], much lower than its rate of uptake would suggest. This is because the ion is transported into the vacuole of the cell, where some 95% of all cell calcium is located. Interaction with calcium-binding proteins such as calmodulin also helps to maintain low intracellular levels. Calcium is also actively excluded from the cell against a concentration gradient across the membrane. Ca^{2+} ions play an important role in the flocculation process.

The relative amounts of calcium and magnesium are known to be critical during yeast growth. For example, increased $Mg^{2+}:Ca^{2+}$ levels in the cell give more rapid initial fermentation rates and higher levels of alcohol production, coupled with higher yeast vitality at the end of fermentation. The converse is true of high $Ca^{2+}:Mg^{2+}$ levels, whence there are extended attenuation periods, mainly due to retarded maltose and maltotriose uptake.

Zinc. Probably the most significant of the trace elements since it can be present in wort in sub-optimal concentrations, especially in breweries where most of the equipment is fabricated from stainless steel. Zn^{2+} plays a critical role in yeast metabolism, and for adequate glycolytic activity 1–2 ppm is the requirement. Brewer's worts can vary in zinc content from 0.1–5.0 ppm, according to how they are prepared, and the element is rapidly taken up by yeast; uptake *via* two systems which are encoded by two genes, ZRT1 and ZRT2. The former controls high affinity uptake which operates in zinc-depleted cells, whilst ZRT2 controls low affinity uptake, characteristic of cells rich in zinc. As with other elements, not all zinc present in wort is necessarily bioavailable; some will be chelated.

At certain concentrations zinc may be inhibitory to growth, the degree of lethality being dependent, amongst other factors, upon the level of Mn^{2+} in the wort. Thus, at wort Mn^{2+} levels of over 0.4 ppm, yeast can tolerate 30 times as much zinc as it can when Mn^{2+} levels are below that figure.

The main zymological significances of the element are: (a) it stimulates maltose and maltotriose uptake; (b) it enhances riboflavin synthesis; (c) it has a stabilising effect on membrane systems; (d) it activates acid and alkaline phosphatases; and (e) it acts as a catalytic centre for several important enzymes, including aldolase, acetaldehyde dehydrogenase and alcohol dehydrogenase (which is a zinc metalloenzyme). Other elements cannot substitute in these reactions. Some zinc has been shown to be stored in the vacuole.

Manganese   Also an essential trace element, being a regulator of several key intracellular enzymes.

Sodium   The yeast does not accumulate Na^+ intracellularly and the cation is continually being excreted from the cytosol in order to maintain very low levels. If cells are subjected to worts of high salt concentration, then they will osmoregulate by producing intracellularly compatible solutes (usually by means of glycerol or arabitol).

Potassium K^+ is taken up actively by yeast, and there is a requirement for glucose, or some other fermentable sugar, in order for this to happen. The uptake of K^+ by the cell is closely associated with the excretion of H^+ from the cell.

During fermentation, yeast exhibits a distinct step-wise affinity for divalent cation uptake, being of the order Mg^{2+}, Co^{2+}, Zn^{2+}, Mn^{2+}, Ni^{2+}, Ca^{2+}, Sr^{2+}.

Oxygen

Overall, oxygen is regarded as being one of he worst enemies of beer; something to be avoided. There is, however, a precise requirement for oxygen by the yeast at the pitching stage. The brewing yeast is unable to synthesise sterols and unsaturated fatty acids in its absence. These two entities are present in wort (malt-derived), but in sub-optimal amounts and so there must be a sufficient level of molecular oxygen in the wort during the early stages of fermentation to facilitate the vast amount membrane synthesis necessary for cell multiplication. As with other factors affecting fermentation, there is a wide range of dissolved oxygen (DO) requirement amongst yeast strains, there being four generally accepted categories:

—Class 1, requiring 4 ppm DO.
—Class 2, requiring 8 ppm DO.
—Class 3, requiring 40 ppm DO.
—Class 4, requiring over 40 ppm DO.

It is now generally accepted that oxygen pick-up during all stages of beer production has a detrimental effect on final flavour, certainly for the lighter, lager-style beers. This even includes the initial stages of brewing, such as milling and mashing,[2] and it has been recommended that oxygen be excluded from all stages of the process, except for yeast pitching. Some breweries even go to the lengths of blanketing mash-mixers and lauter tuns with inert gases such as nitrogen in order to exclude oxygen. Bamforth[3] considers such measures to be 'excessively sophisticated' and that far more simple ways of preventing oxygen ingress can be employed.

YEAST: PHYSIOLOGICAL CONDITION AND VITALITY

There are several methods available for estimating the physiological condition and activity of a yeast sample (this is especially important prior to pitching). Examples are:

(1) measurement of intracellular pH;
(2) rate of oxygen uptake;
(3) rate of CO_2 evolution;
(4) amount of proton extrusion (a sign of membrane integrity);
(5) determination of some intracellular components, such as glycogen, trehalose and sterols (especially ergosterol);
(6) staining methods;
  (a) methylene blue;
  (b) 1-anilino-8-naphthalenesulfonic acid;

(c) rhodamine B;
(d) fluorescein diacetate (FDA);
(e) erythrosin B.

Some of these methods are rather non-specific, whilst others are unrealistic in terms of the time taken to yield results. The brewer needs to know almost immediately what state his yeast is in and how it is likely to perform when it is pitched.

In 1994, Jespersen and Jakobsen[4] used flow-cytometry to estimate cellular events in yeast. Flow-cytometry, using FDA as a stain, provided a rapid method of yeast culture assessment. Cells assume a fluorescence intensity (FI) which varies throughout the course of a fermentation. FDA itself is not fluorescent, but once taken up by a sound yeast cell (there being passive transport across the membrane) it is hydrolysed, by esterases in the cytoplasm, to fluorescein which is then secreted. Secretion is an energy-requiring process and fluorescence emission is dependent upon intracellular pH. Yeast cells exposed to stress conditions show low FI values, but when such cells are placed in fresh growth medium (*i.e.* wort) and relieved from stress, FI values rapidly increase.

One of the important facets nowadays is yeast vitality. This can be defined as the ability of the yeast to endure stress and still 'perform' in the brewery. It represents the physiological state (condition) of a culture and is of prime importance in enabling the brewer to carry out consistent fermentations. In 1996, Axcell and O'Connor-Cox[5] demonstrated that stress tolerance was directly related to fermentation performance, and developed the Stress Test, by which the entire yeast system was evaluated, not just one aspect of it. The test proved to be a more reliable predictor of fermentation performance, but the results took three hours to be produced.

Probably the most rapid, sensitive and reliable test for vitality was developed by Mochaba *et al.* in 1997.[6] These workers had established previously that, upon inoculation into a glucose-containing medium, an alteration in the permeability of the yeast plasmalemma (in preparation for sugar uptake and utilisation) results in the immediate release of low molecular weight compounds, including PO_4^{3-}, K^+ and Mg^{2+}. These ions are later re-absorbed. They also found that yeast slurries giving rise to good fermentations (as judged by shorter lag phase, higher yeast counts, higher levels of ethanol production and lower levels of diacetyl production) released the above ions immediately upon inoculation into wort, whereas cultures giving rise to poor fermentations (the opposite of above) absorbed those ions immediately upon inoculation. From that observation the Magnesium Release Test (MRT) was developed. The MRT provides a rapid means (less than 15 minutes) of assessing yeast vitality prior to pitching and involves spectrophotometry and a readily available colour test kit (Sigma). Methylene blue staining and assessment of protease activity showed that the cells were not dead (autolysed), which would also lead to Mg^{2+} release.

FERMENTATION

Hopped wort that has been run into a fermentation vessel has to be pitched with yeast as soon as possible. The normal pitching rate is of the order of 1 lb of pressed yeast per barrel of wort (0.3 kg hl^{-1}), where small variations cause very little difference in overall fermentation performance and beer flavour. Significant underpitching leads to slow initial fermentations, whilst overpitching (say twice the normal rate) causes undue competition for nutrients, resulting in poorer yeast growth and an ultimate increase in the level of certain esters (*e.g.* ethyl acetate).

Wort temperature at pitching is also important; if wort temperature is more than 5 °C lower than the temperature at which the pitching yeast has been held, then 'cold shock' will result. This manifests itself in a prolonged period of inactivity (*i.e.* extended lag phase).

For the first few hours after pitching, nothing visibly happens in the vessel. This is the lag phase of growth and is an integral part of the growth cycle of any micro-organism that is inoculated into a batch of fresh growth medium. The lag phase can last anywhere from 6 to 15 hours. For industrial fermentations, however, it is important that the desired microbe gets off to as fast a start as possible, thus discouraging the growth of other (unwanted) organisms. This is certainly true for brewery fermentations. Although there are no outward manifestations of metabolic activity, several important physiological and biochemical events are occurring. In effect, the yeast is adjusting itself to the new growth medium (*i.e.* wort), particularly important being the osmotic effect of the sugars contained therein. New enzyme systems are being synthesised to enable the yeast to utilise the wide variety of wort constituents. One such example of this facet of the lag phase is the induction synthesis of a carrier which will permit maltose to enter the cell. The enzyme carrier is maltose permease, and it is produced simultaneously with a maltase [$\alpha(1 \rightarrow 4)$-glucosidase] which enables maltose to be hydrolysed to glucose immediately it enters the cell.

A permease is also required for the uptake of maltotriose. The production of both permeases is inhibited by low levels of glucose and fructose; an example of catabolite repression (see below and page 119). Worts containing significant proportions of glucose syrup or invert sugars exhibit enhanced permease repression, with the consequential slow uptake of maltose and maltotriose. Maltotriose permease production is also inhibited by the presence of maltose.

The passage of maltose and maltotriose across the yeast cell membrane is not a passive process and requires energy in the form of ATP.

If an actively-growing yeast culture (*i.e.* one that has just been skimmed from a vigorous fermentation) is pitched into fresh wort the cells will, in theory, contain all of the enzymes necessary to facilitate uptake and metabolism of all wort constituents. The lag phase will, therefore, be shortened. After a short while, however, the rate of maltose utilisation will decline (due to catabolic repression of maltose permease and α-1,4-glucosidase) until glucose and fructose levels have been sufficiently depleted.

Pitching yeast that has been removed from wort and stored before being re-pitched will have lost the propensity to utilise all but the simplest wort constituents, and so the lag phase will be prolonged whilst re-synthesis of enzymes occurs.

The order of uptake of wort sugars by yeast follows a definite pattern, which, in the main, is governed by the size of the molecule, the concentration of sugar present and the availability of enzyme systems required for metabolism.

Thus, in conventional brewery fermentations, glucose, fructose and sucrose, which are present in low concentrations, are taken up and utilised more or less immediately (sucrose being hydrolysed extracellularly first). This is because the yeast plasmalemma contains a constitutive glucose permease carrier which permits rapid uptake of monosaccharides (glucose first). These sugars will normally be used up from the wort by the second day of fermentation.

Maltose, which is present in far higher concentration, is not taken up until approximately 24 hours have elapsed, and the bulk of it will have been metabolised by day three. Maltotriose, the largest of the major wort sugars, is not normally utilised until the end of the third day of fermentation, when maltose levels are sufficiently low to permit the induction of maltotriose permease. Much of the residual sugar at the end of a normal fermentation is in the form of maltotriose.

Under certain conditions, a yeast may mutate and lose the ability to synthesise maltotriose permease (irrespective of whether there is maltose repression or not). This causes poor wort attenuation, a condition known as a 'sticking' fermentation.

During the early stages of fermentation, it is important that there is sufficient oxygen dissolved in the wort to permit synthesis of membrane sterols and fatty acids. Wort must, therefore, be agitated (roused) or pre-oxygenated. Yeast strains have widely differing oxygen requirements and it is important that there is sufficient oxygen for rapid initial cell growth (especially in terms of membrane constituent synthesis) but not enough for the yeast to undergo aerobic respiration *via* the oxidation and decarboxylation of pyruvate and ultimately the Krebs' cycle. If excessive oxygen is provided, then the products are carbon dioxide and water. The events relating to the Krebs' cycle take place within the mitochondria.

If an early fermentation becomes oxygen-starved for some reason, it is possible to reverse the deficiency by addition of ergosterol (one of the three major yeast membrane sterols) in low concentration (5 µg l^{-1}). Oleic acid can also be used to reverse the situation, which most commonly arises during high-gravity brewing.

Anaerobic conditions are quickly established in the fermenter and yeast readily converts fermentable sugars (ultimately glucose) to ethanol and carbon dioxide. This is effected through the glycolytic or Embden–Meyerhof–Parnas (EMP) pathway (Figure 5.4).

This is the main energy-generating process for the yeast, the oxidative part of the process resulting in the generation of the reduced coenzyme NADH. NADH is re-oxidised when acetaldehyde is reduced to ethanol. Obviously, some carbon from glycolysis is required for yeast growth (biosynthesis) *via* pyruvate dehydrogenase. This results in a depletion of the NAD^+ pool, a situation that will drastically affect the redox balance in the cell. The situation is compensated for, however, by a series of redox-balancing mechanisms such as the production of glycerol, the terminal steps in the production of higher (fusel) alcohols and the reduction of diacetyl.

It has been known since 1929 that at high glucose concentrations (>0.4%), even under aerobic conditions, *Saccharomyces cerevisiae* metabolism is fermentative rather than oxidative. This is associated with morphological and biochemical alterations in the mitochondria, which resemble those in cells grown under anaerobic conditions. Biochemically, the most significant differences are the deficiency of certain Krebs' cycle enzymes, particularly succinic and α-oxoglutarate dehydrogenases, and some respiratory chain components, notably cytochromes. This is known as the Crabtree Effect, the Reverse-Pasteur Effect, or Glucose Effect. It is another example of catabolite

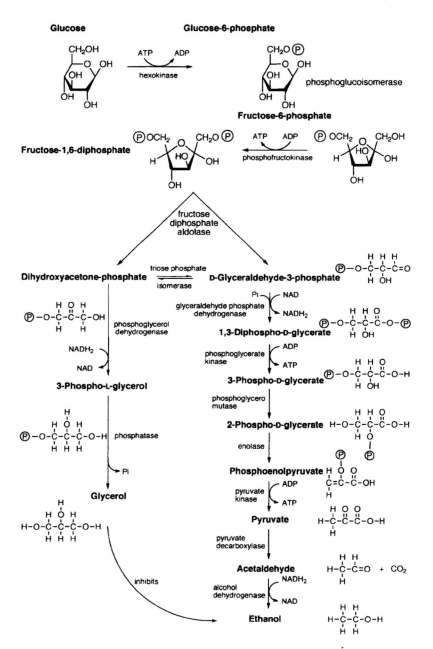

Figure 5.4   *The Embden–Meyerhof–Parnas (glycolytic) pathway*

repression[7] and is the prime reason for determining the course and nature of brewery fermentations, which, in essence, are all under the regulation of the Crabtree effect.

The Pasteur effect describes the deleterious influence of oxygen on fermentation. In 1861, Pasteur found that fermentation in yeast was inhibited by the presence of oxygen, *i.e.* when yeast was transferred from an anaerobic to an aerobic environment, growth was accelerated while uptake of sugar was diminished. The latter observation is, in part, attributable to the difference in the Michaelis constant (K_M value) for the accumulation of glucose under the two sets of conditions, as seen in experiments by Kotyk and Kleinzeller[8] in 1967. They found that when yeast was grown anaerobically the K_M value for glucose uptake was 6.7 mM, whilst under aerobic conditions it is of the order of 17.4 mM. These observations were linked to the affinity of glucose for its permease, which is decreased in the presence of oxygen. The enhancement of vegetative growth (cell material) is simply explained by the fact that *Saccharomyces cerevisiae* can produce more ATP from each molecule of glucose under aerobic conditions. Theoretically, fermentation should proceed for as long as anaerobic conditions prevail and there being a source of hexose sugar in the wort. Under the conditions of brewery fermentations, however, this does not happen. There are several reasons for this:

(1) The growth of yeast cells during active fermentation is by budding, a process that requires plentiful cell wall and membrane synthesis. Because of the prevailing anaerobic conditions, sterol and fatty acid synthesis cannot occur and thus membrane formation declines and a point is reached whereby growth ceases.

(2) As fermentation proceeds, the level of ethanol increases to such an extent that it becomes toxic to yeast growth.

(3) With the slowing down of yeast growth in the fermenter, conditions become less turbulent, particularly with respect to CO_2 evolution. This causes yeast cells to flocculate and, hence, sediment out at a faster rate. This enhances cessation of yeast growth.

The lag phase is followed by a short phase of accelerating growth (Figure 5.5), which leads to a phase of exponential growth. During this logarithmic period of growth the yeast density (bulk) is expected to increase by four- to six-fold. It is at this stage that cell growth is at

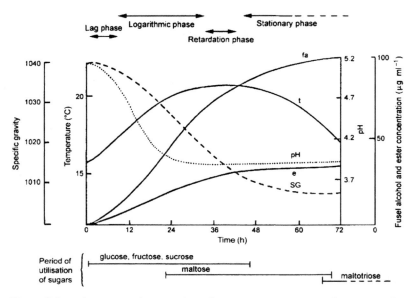

Figure 5.5   *The course of a typical top fermentation:* SG = *specific gravity of wort;* t = *temperature;* fa = *fusel alcohols;* e = esters

its highest level, with cells multiplying by budding and rapidly producing ethanol and carbon dioxide. A considerable amount of heat is produced and attemperation is normally necessary in order to keep the temperature down to 20–22 °C (for ale fermentations). If the temperature rises above this level, other higher alcohols will be produced (such as propanol, iso-amyl alcohol, iso-butyl alcohol) which impart off-flavours. Logarithmic growth normally persists for 48–60 hours, after which a phase of decelerating growth (retardation phase) is entered before the cells reach the stationary phase. The latter signifies the end of primary fermentation *per se*. A small number of new cells are being produced during the stationary phase, but they are counteracted by the number becoming moribund.

Yeast required for subsequent fermentations should be skimmed from the top of the fermenter at the end of the exponential phase. Cells recovered later on during fermentation will be less viable and more likely to contain contaminating organisms.

In addition to ethanol and CO_2, there are a number of minor products of fermentation (and of yeast sugar-metabolism in general). These may be put into four major categories:

(1) *Fusel alcohols.* These are higher alcohols whose production is largely governed by the amino acid composition of the wort, as long as there is sufficient sugar available. There are two pathways by which higher alcohols are produced:

(a) The Ehrlich pathway, whereby amino acids are taken up from wort and transaminated in the cell to α-keto acids. These acids are subsequently decarboxylated to aldehydes, which are reduced to yield alcohols with one fewer carbon atom than in the starting amino acid. High levels of an amino acid (in excess of that required for routine cellular metabolism) lead to enhanced levels of the fusel alcohol analogue. Thus, for example, excess valine would promote the production of isobutanol during fermentation; the pathway proceeding *via* the oxo-acid, α-oxoisovaleric acid and the aldehyde isobutyraldehyde.

(b) The normal amino acid biosynthetic pathway (*via* aminated carbon skeletons). The penultimate step in these pathways is the formation of an α-keto acid which is finally transaminated to an amino acid. If the α-keto acids are decarboxylated and reduced (as happens in the Ehrlich pathway) then higher alcohols are liberated.

In both pathways the final reduction of aldehyde to alcohol involves NADH (with the production of NAD^+), and it is thought that the *raison d'être* for higher alcohol production by yeast is partly aimed at maintaining redox balance. There may also be an attempt to regulate intracellular pH, since α-keto acids are being removed from the cell.

A generalised scheme for the production of fusel alcohols is given in Figure 5.6.

Other factors influencing the production of higher alcohols are:

—elevated fermentation temperatures;
—excessive rousing;
—excessive yeast growth;
—the strain of yeast itself.

The main higher alcohols found in beer are propanol, iso-butanol (2-methyl propanol), *n*-butanol, iso-amyl alcohol

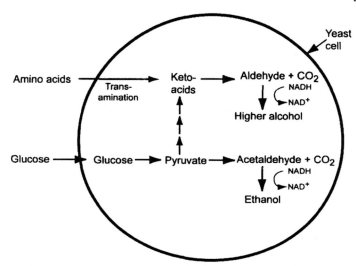

Figure 5.6 *Generalised scheme for the production of higher alcohols*

(3-methyl butanol) and *n*-amyl alcohol (2-methyl butanol). All of these have an influence on beer flavour. Ethanol itself makes little contribution to overall beer flavour, but it does have a considerable influence upon the flavour delivery of other beer constituents.

(2) *Acids.* A wide variety of organic acids are produced during fermentation, some of which make important contributions to flavour. There are two groups:

    (a) Volatile organic acids; principally acetic, propionic and butyric acids. The former is produced by the hydrolysis of acetyl CoA.

    (b) Non-volatile organic acids. These include the oxo-acids, such as pyruvic acid, and products of the Krebs cycle, such as succinic acid. Oxo-acid production is, again, closely linked to the amino acid composition of wort.

(3) *Esters.* These are the result of the reaction between alcohols and acids, and about 80 have been identified in beer, ranging from C_3 (*e.g.* ethyl acetate) to C_{17} (*e.g.* 3-methylbutyl dodecanoate), although, given the number of alcohols and acids present in beer, there could theoretically be around 3700 of them.

    Esters are of prime importance in influencing beer flavour, particularly if found in excessive quantity where they lead to 'off-flavours'. Some have extremely low flavour thresholds [*e.g.* ethyl butyrate at 0.2 ppm (mg l^{-1})].

Because of the preponderance of ethanol in beer, ethyl esters are by far the most commonly encountered, with ethyl acetate being found in the highest concentration [normally from 10–70 ppm (mg l^{-1}) depending upon beer style]. Wild yeasts, such as *Hansenula* and *Pichia* species, can produce high levels of ethyl acetate *via* an aerobic pathway if they are present as contaminants in sufficient numbers.

Since acetic acid is a by-product of fermentation, there is no reason why ethyl acetate cannot be formed by direct esterification with ethanol. It has been shown, however, that the quatities present in beer are far higher than those that could be accounted for by the slow, direct combination of alcohol and acid. The reaction must, therefore, be under enzymic control, particularly since it is known that the ability to esterify is a stable genetic property of brewing yeasts. Esterases have been located in *Saccharomyces cerevisiae*, but it seems that they do not play a significant role in ester production. The reason for proposing this is that when (under laboratory conditions) the rate-limiting (lowest concentration) component in the reaction (*i.e.* acetic acid) is added to the reaction mixture, no increase in ethyl acetate production is observed.

In 1962, Nordstrom[9] showed that ester formation could be enhanced by artificial addition of pantothenic acid, which is a precursor of coenzyme A (CoASH). This finding led to the establishment of the fact that acyl coenzyme A compounds (in particular, acetyl CoA which is most abundant) play an active role in esterification during fermentation. Suppression of yeast growth, by whatever means, results in the formation of inordinately high levels of acetate esters; at the expense of other acyl esters. This is purely because of the key role played by acetyl CoA in metabolism; there is always sufficient present to enable acetate esterification to occur.

In yeast, acyl CoA compounds are formed in several ways: from keto acids, from fatty acid biosynthesis and from fat catabolism.

Reaction (5.1), for the formation of ethyl acetate, is as follows:

$$C_2H_5OH + CH_3COSCoA \longrightarrow C_2H_5CO_2CH_3 + CoASH \quad (5.1)$$

The reaction is catalysed by the enzyme acyl-alcohol transferase (AAT).

Direct esterification does occur, especially during secondary fermentation, whilst a few esters are derived from the volatile fraction of hops. Some derived from the latter source undergo transesterification with those produced during fermentation.

A number of γ-lactones are produced in small quantity during fermentation. These are cyclic esters which originate from hydroxy acids.

Generally speaking, ester synthesis is a very complex phenomenon, there being no discernible direct relationship between yeast growth and ester synthesis.

(4) *Aldehydes and ketones.* Both of these groups of carbonyl compound are found in beer in low concentration, but only aldehydes are produced to any extent during fermentation. The ketones identified in beers, some 20 of them, are very minor products of fermentation and mostly originate from the degradation of hop products.

Approximately 30 aldehydes are known from the analysis of various beers, ranging from C_2 (acetaldehyde) to C_{12} (dodecanal). Acetaldehyde is found in the highest concentration $(4-15 \text{ mg l}^{-1})$, which is to be expected since ethanol is formed during fermentation by the reduction of this compound. The formation of all aldehydes follows the same principle, each being synthesised by the enzymic decarboxylation of the relevant α-keto acid. Thus, acetaldehyde arises from the decarboxylation of pyruvic acid (pyruvates) by pyruvate decarboxylase.

Each conversion of α-keto acid to aldehyde requires a specific decarboxylase enzyme, but the reduction of aldehydes to their respective alcohols (a process which oxidises NADH to NAD^+) employs a single, non-specific, alcohol dehydrogenase.

Most aldehydes appear to arrive in beer by small-scale 'leakage' before they are reduced to alcohols, but a fraction of them arises from subsequent oxidation of alcohols during secondary fermentation and conditioning.

One of the most flavour-active compounds produced during fermentation is diacetyl, which confers a toffee or butterscotch taste. It is formed indirectly as a by-product of the biosynthesis of valine in the cell. The key compound is the acetohydroxy acid, α-acetolactate (AAL), an intermediate in valine synthesis. This compound leaks out of the yeast cell and is then non-biologically converted to diacetyl in the wort, the reaction being an oxidative decarboxylation. The mechanism occurs entirely independently of yeast and is favoured by low pH and elevated temperatures. Molecular oxygen is not a prerequisite, but an (as yet unknown) oxidising agent is known to operate. Diacetyl then re-enters the cell whence it is reduced to acetoin and 2,3-butanediol, reactions involving NADH. As a result, oxidised NAD^+ is produced and, again, this is thought to be important in maintaining the redox balance in the cell during the latter stages of fermentation, when diacetyl is normally taken up. Acetoin and 2,3-butanediol have far higher taste thresholds and, therefore, do not contribute such adverse flavours to beer.

Another acetohydroxy acid, α-acetohydroxy butyrate, also produced by yeast during fermentation, is similarly extruded by leakage and is extraneously oxidised, this time to the diketone 2,3-pentanedione. Whereas α-acetolactate emanates enzymatically from pyruvate in the cell, α-acetohydroxy butyrate is a product of α-oxo butyrate.

Diacetyl and 2,3-pentanedione are, together, categorised as vicinal diketones, and a generalised scheme for the production of diacetyl during fermentation is given in Figure 5.7. Flocculated yeast cannot remove vicinal diketones.

The deleterious flavour effects of diacetyl are far more

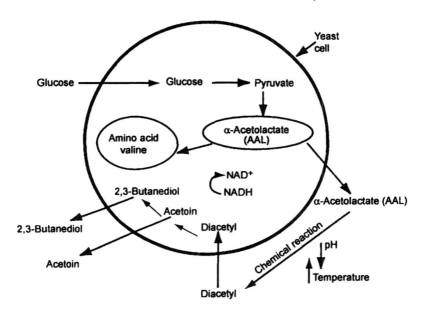

Figure 5.7 *Scheme for the formation of diacetyl*

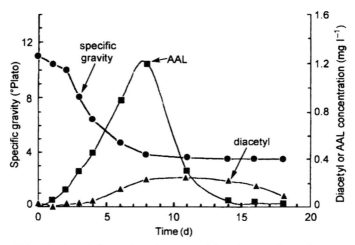

Figure 5.8   *AAL and diacetyl production and fate during a lager fermentation*
(Courtesy of the *Brewers' Guardian*)

pronounced in pale, lager-style beers, and its flavour threshold in such beers can be of the order of 15–$20 \mu g l^{-1}$.

The timescale for the production of AAL and diacetyl during a normal lager fermentation is shown in Figure 5.8.

Glycerol is produced during the early stages of fermentation (see Figure 5.4) as a result of a mechanism to prevent the build-up of NADH (caused by the rapid metabolism of pyruvate). The yeast uses NADH to reduce dihydroxyacetone phosphate to glycerol phosphate, which is subsequently dephosphorylated. Much of it is used for internal cell syntheses (*e.g.* phospholipids and triglycerides) but some is secreted and can be present in certain beers up to a level of $3.5 \, g \, l^{-1}$.

Under aerobic conditions, yeast can use glycerol as a source of carbon.

Storage Compounds

Towards the end of the fermentation cycle two intracellular storage carbohydrates are produced by the yeast; glycogen and trehalose. Both are composed of glucose units; glycogen is polymeric, trehalose is dimeric. Their synthesis originates with the formation of uridine diphosphate glucose (UDPG). Trehalose is a non-reducing disaccharide, the two glucose units being α-1,1-linked through their reducing carbon atoms. Unlike glycogen, trehalose may also be synthesised from glucose-6-phosphate. Trehalose is located in the cytoplasm and becomes closely associated with the plasmalemma under conditions of stress (such as starvation or heat-shock). It is now considered, by some authorities, to be a stress protectant rather than a food reserve. One can now use the the trehalose content of the cell as an indicator of yeast vitality and viability, thanks to a rapid detection method using near infrared (NIR) spectroscopy. One can also use NIR to detect glycogen.

Fermentation Systems

As with the controlled cultivation of any micro-organism, the growth of yeast for beer production can occur on a batch basis, or it can be a continuous process.

Batch Fermentation

The earliest documented fermentations were all on a batch scale, *i.e.* a predetermined volume of wort is fermented and a corresponding volume of beer is produced.

Traditionally, fermentation was carried out in open, rectangular vessels constructed of metal (usually copper), wood, stone or slate. Stainless steel is now the most often used material, although synthetic polymers have found favour in certain circumstances. The design of fermentation vessel has changed over the years, always with the aim of increasing efficiency, hygiene and the likelihood of more consistent beer. Perhaps the ultimate in design and technology are the continuous fermentation systems which were the vogue in the 1960s, and which are still used in some breweries today. Some systems have been devised to accommodate certain water and/or yeast peculiarities encountered in some brewing centres; the Burton Union method of fermentation is a classic example (now only used by Marston, Thompson & Evershed in Burton-upon-Trent) (see page 130).

Regardless of shape, the major features of an open fermentation vessel suitable for the production of traditional ale are that it should have a flat bottom, a beer extraction port at, or somewhere near to, the base, and some form of attemperation. The latter may be achieved by re-circulating chilled water through a coiled ring system inserted inside the vessel (Figure 5.9), or by means of an external cooling jacket. The latter arrangement is preferable because there are no associated cleaning problems. A considerable amount of heat is

Figure 5.9   *View into a fermentation vessel showing internal attemperation*
rings (Nethergate Brewery)

generated during the fermentation of sugars to ethanol and CO_2, and
if the yeast is allowed to metabolise at elevated temperatures, undesirable by-products can result. There is also a greater danger of infection
at temperatures much above 22 °C. Most larger breweries have their
fermentation vessels situated in temperature-controlled rooms.

It is essential that vigorous, top-fermenting yeasts are used in open
fermentation vessels. The yeast head that forms on top of the
fermenting beer (Figure 5.10) together with the overlying zone of CO_2
provide a barrier to airborne contaminants and keep oxygen away
from the green beer. At a certain stage of the fermentation cycle, yeast
has to be skimmed from the top of the vessel. Some of this will be used
for subsequent fermentations. There are several ways of yeast-skim-
ming, one of the more modern being by vacuuming off. Beer produced
in open fermentation vessels is supposed to possess a characteristic
'clean palate', according to beer connoisseurs.

Apart from any likelihood of infection, the other drawbacks of
open fermenters are an inability to harvest and re-use the CO_2
produced, relatively high beer losses when skimming the yeast (during
and after fermentation), and that vessels normally have to be hand-
cleaned.

A variation of the standard open fermenter was developed in

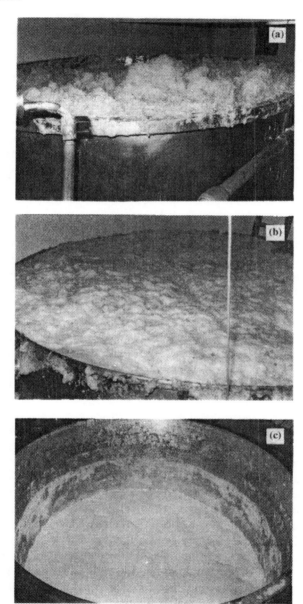

Figure 5.10 *Fermentation: (a) ca. 12 hours, showing 'cauliflower heads'; (b) 24 hours; (c) 72 hours, the yeast has been stripped from the top and the remaining yeast head has compacted*

certain Yorkshire breweries that used extremely flocculent yeasts. The vessels were called Yorkshire Squares and were constructed of stone or slate. Each consists of two compartments, one overlying the other. The upper and lower compartments are separated by a platform, and the two are linked by a manhole, some 60 cm in diameter, and a series of pipes. Wort is introduced into the lower chamber, which is completely filled, and into the upper chamber to a depth of approximately one inch. During fermentation the yeast rises through the manway into the upper chamber, whence it can be skimmed. The whole fermentation has to be periodically pump-roused in order to re-circulate the yeast which would otherwise sit in the upper compartment. After fermentation is complete, the beer is drawn off from the lower compartment. Again, beer losses are quite heavy and cleaning is very labour-intensive.

Some of the Midlands breweries, especially in the Burton-upon-Trent area, employed non-flocculent yeast strains which posed problems if used in standard open fermenters. This led to the development of a unique fermentation apparatus called the Burton Union system. The unions themselves are constructed of 24 wooden casks (in two rows of 12), each of about five barrels' capacity. The casks are fitted with internal cooling rings and, at the top, have a protruding swan-neck pipe. A series of gulleys, or troughs lies above and below the casks, the upper trough being slightly inclined. A series of pipes emanates from the lower end of the upper trough; these lead back to the ends of the rows of casks and are called side rods.

Chilled bitter worts are introduced into a collecting vessel where yeast is pitched and fermentation commences. After about 36 hours the actively fermenting beer is dropped into the union casks. Continued fermentation causes beer and yeast to be forced out of the swan necks and discharged into the upper trough. Beer flows slowly down the incline and ultimately returns to the casks *via* the side rods. Yeast tends to settle out in the upper trough after the beer has trickled away. By the time fermentation has been completed most of the yeast will have been separated from the beer in the union casks. The beer is then dropped into the lower trough whence it is carried to a racking vessel. Beers produced by the Union method have totally unique and highly-regarded flavours. As would be expected from the very nature of the Burton Union systems, they are difficult to clean and beer losses can be very high.

Closed fermenters were first used before the 1914–18 war, but they were not familiar pieces of equipment in the brewing industry until the 1960s. The earliest patent for an enclosed vessel was by L. Nathan in

1908; he followed this with a further design in 1927. The original examples were constructed from aluminium and were vertical, cylindrical vessels with conical bottoms (*i.e.* cylindroconical vessels). Nowadays, stainless steel is the usual construction material. Apart from the obvious advantage of the elimination of airborne infection, a number of other claims have been made for this type of vessel (which is often referred to as a Nathan tank):

—greatly reduced beer losses;
—CO_2 can be collected;
—easy recovery of yeast;
—can be cleaned automatically;
—can be used for conditioning as well as fermentation;
—their height to width ratio allows more effective use of floor storage space in the brewery;
—reduced fermentation times.

Cylindroconical vessels are equipped with an external cooling jacket and are usually lagged in some way, often with fibreglass. Vessels normally have a height to diameter ratio of approximately 4 : 1 and, when primary fermentation is proceeding, strong circulatory currents are produced due to vigorous CO_2 production, particularly near the base of the vessel where the bulk of the yeast is situated. This, in effect, leads to a 'self-rousing' mechanism and greatly enhances the rate of fermentation. It is claimed that fermentation times can be reduced by as much as 50%. It is also claimed that beer produced in cylindroconical fermenters show more efficient utilisation of bitter substances (none being lost with the yeast head) and greater powers of head retention. The CO_2 levels of beers fermented in these vessels are far higher than those produced from any other batch fermentation system.

When the first cylindroconicals were installed in existing breweries, their size was disproportionally large as compared with the rest of the brewhouse (mash tun, copper, *etc.*). In some cases, this necessitated several brewlengths to be run before the fermenter became filled with wort. If care was not taken, this could lead to problems because different the batches of wort would be at variable stages of oxygenation and fermentation. Wort-filling of cylindroconicals should be no more than 75% of the total volume of the vessel; this leaves enough head space for the copious amount of foaming produced during fermentation.

The yeast used in breweries that employ cylindroconicals will be a bottom-cropping strain of *Saccharomyces cerevisiae*. At the end of primary fermentation, the yeast sediments out in the conical bottom of the vessel, where it can be aseptically withdrawn.

An enormous amount of work has been carried out into the design of these vessels, particularly with regard to the location of cooling jackets and the slope of the conical base. In respect of the latter, it has been shown that an included angle of 70° is the optimum for ensuring maximum yeast settlement. Cylindroconical vessel sizes can vary from 60 to 7000 barrel capacity. A dual-purpose variation of the Nathan tank, called the 'Unitank', has been developed by the Rainier company. They have a shallow (no more than 25° off horizontal) conical bottom and are used for primary fermentation and conditioning.

Continuous Fermentation

The notion of using continuous fermentation for beer production was first forwarded by Max Delbruck in 1892.[10] The first British patent for such a means of fermentation was taken out by L.A. van Rijn in 1906 (British Patent No. 18 045), but very few breweries adopted the process until the 1960s.

Many advantageous claims have been made for these fermenters, most of which have not been substantiated by practical experience. The main supposed advantage was increased efficiency, particularly in terms of labour input and plant utilisation. This has not been found to be the case, labour costs in particular being similar to more traditional methods of fermentation. Cleaning certainly can be carried out with greater ease – as long as no serious infection problem arises! The claim for lower beer losses during production also cannot be justified, losses being the same as for cylindroconical fermenters. It was also hoped that the continuous nature of the fermentations would lead to more consistent beer, but this is only possible if there is a consistent, continuous production of hopped-wort. One certain advantage is the enormous reduction in primary fermentation time, which, under certain conditions, can be reduced to 6–8 hours. It is also possible to run the process for extended periods of time (months, if required) without cleaning and re-charging with yeast. This is provided, of course, that there is no outbreak of infection.

The efficiency of transforming fermentable sugars into ethanol and CO_2 is much higher under continuous conditions. This is principally because, in this greatly modified physical environment, the yeast does not convert fermentables into cell constituents at a very rapid rate. Thus, there is more concentration on fermentation *per se*.

Of the two continuous processes that have been employed to any extent, the 'open', or cascade system is less complicated and cheaper to instal. The apparatus usually consists of three main vessels; two for fermentation, and one to act as a yeast separator, although more fermenters may be used in some breweries. Cooled, sterilised hopped-wort is pumped through an oxygenater into the base of the first fermenter and the yeast is introduced. This is then constantly stirred to maintain homogeneity. Some 50% of the required fermentation occurs at this stage. The partially fermented wort is then run into the second fermenter, which is under totally anaerobic conditions; fermentation is completed here. The yeast concentration in the first fermenter is held in a steady state owing to the fact that the number of cells being produced is counteracted by the number lost to the second vessel. Beer and any suspended yeast drop into the third vesel, which is a sedimentation tank that is equipped with cooling rings and a conical bottom. Chilling in this vessel promotes the settling of yeast into the bottom cone, whilst beer is drawn off from a higher level for conditioning and processing.

New Brewing Yeasts

Ultimately, it is the yeast that determines the pattern of fermentation in the brewery. Most brewing strains have the following inherent limitations:

—susceptibility to contamination; especially when open fermentation vessels are used;
—limited fermentation efficiency;
—limited range of carbohydrates fermented;
—limited tolerance to alcohol, temperature and CO_2.

Conventional genetic techniques, whilst producing many new laboratory strains of *Saccharomyces cerevisiae*, have not yielded any new brewing strains. The original techniques involved – mutagenesis, mating and protoplast fusion – all failed to liberate genetically stable brewing strains. The reasons may be summarised as follows:

(1) *Mutagenesis*. Genes are disrupted by exposure to a physical (*e.g.* ultraviolet light) or chemical mutagenic agent (*e.g.* ethyl methanesulfonate). This is very non-specific and other genes may be altered as well as the desired one. There is always the likelihood of the mutated gene reverting in the course of time;

instability is a major problem. One cannot impart new properties by mutagenesis; one is merely changing the expression of an
existing gene.

(2) *Mating*. Brewing strains are mostly polyploid and do not
sporulate readily. Thus, the haploid cells needed for mating are
difficult to produce. In addition, mating rates are poor.

(3) *Protoplast fusion*. The low mating rates encountered in brewing
strains are partly attributable to the cell wall recognition sites. If
cell walls are enzymatically removed, protoplasts are liberated
and these can be manipulated and fused; one being the donor
protoplast, the other the recipient. Their nuclei will then fuse
and diploidy is enstated. Cell walls subsequently regenerate.
The main problem is that only 'compatible' protoplasts can fuse
and, even when this happens, one of the nuclei tends to
'dominate'.

It is now generally accepted that recombinant DNA techniques are
the only way to obtain the required control and specificity in yeast
manipulation. Much of the original work was carried out at the BRFI
(BRI) and some new, useful properties have been introduced into
brewing strains. Pilot-scale fermentations have been carried out and a
number of acceptable beers produced.

Recombinant DNA techniques have been employed to attempt to
impart the following desirable characters into brewing strains of yeast:

(1) utilisation of wort dextrins;
(2) removal of wort β-glucans (viscosity and haze improvement in
    beer);
(3) controlled flocculation;
(4) rapid diacetyl removal (reduced maturation times);
(5) resistance to bacterial contamination;
(6) resistance to killer toxin;
(7) utilisation of pentose sugars.

A summary of some of the developments in each of the above seven
categories is as follows:

(1) One of the most successful yeast modifications has seen the
    incorporation of starch-degrading (amylolytic) enzymes into the
    genome. By the very nature of normal brewery wort production
    and fermentation, most beers contain dextrins which are not
    fermented by *Saccharomyces cerevisiae* or *Sacch. uvarum*. They

contribute to sweetness and calorific value. *Sacch. diastaticus*, which is related to brewing yeasts, can partially break down dextrins because it possesses the enzyme glucoamylase, which converts them to fermentable glucose. The breakdown is not complete, however, because the glucoamylase does not have any de-branching activity, only the dextrin backbones are fragmented. *Sacch. diastaticus* glucoamylase has been inserted into brewing strains and some very well attenuated beers have been produced. Unfortunately, these strains are genetically unstable and fermentations proceed at a slow rate. Glucoamylase from *Schwanniomyces occidentalis* and *Aspergillus niger*, however, does show de-branching activity and wort dextrins can be completely degraded. Brewer's yeasts containing the *Schwann. occidentalis* enzyme do show normal fermentation patterns and produce beers of very low calorific value.

Extensive trials at BRFI (BRI), with a lager yeast (NCYC 1342) transformed with three copies (in three identical chromosomes) of *A. niger* glucoamylase, have resulted in the production of a standard low-calorie beer called *Nutfield Lyte*. The yeast has proved to be stable.

The grist consisted of 80% malt and 20% high maltose syrup and the worts are collected at 1044°. Fermentation is carried out for 12 days at 12 °C. Super-attenuation and flavour maturation commence after day five. After 12 days the beer is attenuated at 998° and the alcohol content is 6.2% (v/v). This is diluted to 5.0% alcohol before packaging.

(2) It is inevitable that a proportion of barley cell wall β-glucan will survive the mashing regime and proceed to the finished beer. Excessive residues of β-glucan can lead to wort run-off and filtration difficulties because of increased viscosity. There are also inherent haze problems. The gene for the glucanase enzyme (*endo*-1,3-1,4-D-glucanase) has been isolated from barley and from certain fungi and bacteria (particularly *Bacillus*).

Bacterial and fungal β-glucanase genes have been successfully transmitted to brewing yeasts and the recipient organisms have secreted the enzyme in sufficient quantity that the trial beers produced exhibited reduced β-glucan levels, and, hence, viscosity. There is apparently little effect on fermentation performance and flavour character of the resulting beer.

(3) Flocculation is the aggregation of yeast cells into large clumps (flocs), ideally at the end of fermentation. Changes in flocculation character can affect fermentation; these are often caused by

a wort aberration and are normally transient but they can be permanent, especially if the yeast has been stored away from wort for prolonged periods. Flocculation is genetically controlled; at least 12 genes are involved, of which FLO_1 is the dominant one. FLO_1 has been introduced, with some success, into non-flocculent strains. The actual mechanism of flocculation involves cell wall proteins (lectins) which bind onto carbohydrate receptors (mannans). Divalent cations (Ca^{2+} or Mn^{2+}) are required for the integrity of lectins.

It has also been established that mitochondrial DNA plays a role, because repiratory-deficient mutants of brewing strains do not flocculate. Flocculation is inhibited by mannose, sucrose, maltose, glucose, NH_4^+ and basic amino acids.

(4) Diacetyl, as we have said, is one of the most important flavour-active compounds in beer (butterscotch or toffee flavour), it being formed during fermentation from α-acetolactate, an intermediate in amino acid biosynthesis. During the maturation (conditioning) of beer the yeast converts diacetyl to acetoin and 2,3-butanediol, which are not flavour-active.

Diacetyl formation is a relatively slow process; diacetyl removal is more rapid. There is much interest in the breeding of yeasts that produce less diacetyl during primary fermentation. There are two main approaches:

(a) To prevent diacetyl formation altogether. The enzyme α-acetolactate dehydrogenase converts α-acetolactate to acetoin directly, completely circumventing diacetyl production. The gene for the enzyme has been inserted into the brewing yeast, and fermentations carried out with this modification have produced beers with minimal levels of diacetyl. Experiments have been carried out on a pilot scale in Finland and Japan.

(b) To promote the activity of the natural α-acetolactate-degrading enzyme in the yeast. The number of genes controlling the production of this enzyme have been increased by manipulation, and as a result, beers with up to 80% less diacetyl have been brewed. At the pilot stage there seems to be little detrimental effect on other aspects of primary fermentation.

(5) A wide range of potentially contaminating bacteria can be found in beer, so it is unlikely that one agent can be found to eradicate them all. One cannot use conventional antibiotics because of the danger of producing resistant strains through

their constant use. Bacteriocins, however, which are not in medical use, have been investigated; these are proteins, secreted by some microbes that are lethal to certain others.

One of the most likely candidates for investigation is nisin, a polypeptide (34 amino acids; mol. wt. 3510 Daltons) produced by the dairy organism *Streptococcus lactis*. Nisin has been used for over 50 years in the food industry as an internationally accepted food preservative.[11] Nisin production is plasmid-encoded and the gene has been cloned and sequenced. Attempts have been made to insert the gene into yeast, but no vector site for its expression is yet known. Also, the lethal action of nisin is on the cell membrane of susceptible organisms and so, unless the nisin produced by the yeast was contained in discrete vesicles upon expression, the yeast membrane will be disrupted as well. Work is still being carried out.

Also being looked at are reuterin and oleuropein. Reuterin is a low molecular weight, water-soluble product of glycerol fermentation by *Lactobacillus reuteri*. It is a broad-spectrum compound naturally found in the human gut. Oleuropein is a phenolic glycoside produced by fresh olives. It is active against a wide range of bacteria but is not deleterious to yeasts.

(6) Occasionally, brewery fermentations become infected with killer strains of yeast; this can occur in batch and continuous fermentations. The resultant beer has to be destroyed because of its phenolic/herbal flavours and the plant must also be scrupulously cleaned. The killer strains have no known effect on humans.

Killer strains are wild yeasts (*Saccharomyces cerevisiae*) that produce a proteinaceous extracellular toxin which binds with the brewery yeast cell wall and forms channels in the underlying plasmalemma. Essential ions and ATP escape through the damaged membrane and the cell loses its integrity and dies. The killer strains are, themselves, immune to the toxin and so attempts have been made to introduce their resistance factor into brewing yeasts with the hope that the 'killer brewing strain' would be able to suppress the growth of wild *Saccharomyces* spp. during fermentation. This has been achieved by conventional mating; a rare example of such a genetic exercise. This is because the killer character is not coded for by the yeast chromosome but by virus-like particles which can be readily transferred from cell to cell. Brewing characteristics are not affected, but such strains have not yet been used commercially. A drawback is the fact that the killer toxin is only effective over a narrow pH range (4.2–4.8). The killer toxin from the yeast *Kluveromyces* has a wider pH range of activity and is also active against a wider range of wild yeasts (including non-*Saccharomyces* spp.).

(7) There are very low levels of pentose sugars in normal brewery worts, but they are a major component (especially xylose) of secondary worts produced by acid hydrolysis of pentosan-rich spent grains. Brewing yeasts cannot use pentose sugars, but if xylose isomerase from *Escherichia coli* can be inserted then the yeast would have the ability to convert xylose to xylulose, which it can utilise *via* the pentose phosphate pathway or the EMP pathway. Should this prove to be successful then it would be possible to ferment secondary worts.

REFERENCES

1 E.M.R. Rees and G.G. Stewart, *J. Inst. Brew.*, 1997, **103**, 287.
2 J.W.M. van Waesberghe and W.J.J.M. van Waesberghe, 'European Brewery Convention Monograph XXIII, Symposium on Malting Technology', 1994, p. 44.
3 C.W. Bamforth, *The Brewer*, 1996, **82**, 22.
4 L. Jespersen and M. Jakobsen, *J. Inst. Brew.*, 1994, **100**, 399.
5 B.C. Axcell and E.S.C. O'Connor-Cox, 'Proceedings of the Institute of Brewing (Asia Pacific Section) Convention', Singapore, 1996, p. 64.
6 F.M. Mochaba, E.S.C. O'Connor-Cox and B.C. Axcell, *J. Inst. Brew*, 1997, **103**, 99.
7 R. Magasanik, *Cold Spring Harb. Symp. Quant. Biol.*, 1961, **26**, 249.
8 A. Kotyk and A. Kleinzeller, *Biochim. Biophys. Acta*, 1967, **135**, 106.
9 K. Nordstrom, *J. Inst. Brew.*, 1962, **68**, 398.
10 M. Delbruck, *Wochshr. Brau.*, 1892, **9**, 695.
11 K. Ogden, M.J. Waites and J.R.M. Hammond, *J. Inst. Brew.*, 1988, **94**, 233.

Chapter 6

Beer – Post-fermentation

Once a beer has been fermented down to its final gravity it is chilled to 9–11 °C. This can either take place in the fermentation vessel or after removal to a subsequent vessel. Beer produced from a primary fermentation is referred to as 'green beer' and this must be conditioned before packaging. The conditioning process can, for certain styles of beer, represent the longest single part of the brewing cycle (and the equipment needed represents up to 25% of the total capital cost of the brewery). There are five main processes to be considered during conditioning:

(1) *To effect flavour maturation.* This involves the removal, by yeast, of some of the undesirable by-products of primary fermentation; most notably, sulfur compounds (including H_2S), acetaldehyde and diacetyl. Removal of the undesirables, a process called purging, is naturally slow and requires the yeast to be in a relatively good metabolic condition.

(2) *To clarify the beer.* Green beer taken off of the fermenter can contain anything up to 1×10^7 yeast cells per ml, which, for brewery-processed beers that involve filtration, could cause severe blockage problems. Thus, a period of storage is necessary to allow some removal of yeast by sedimentation. The yeast must not be taken away from the beer too quickly, otherwise the maturation processes mentioned above will not take place. Yeast settlement can be promoted by the addition of clarification aids (such as finings), although this is not permitted in certain breweries, especially those conforming to the German Purity Law.

(3) *To stabilise the beer.* Beers that are required to have an extended shelf-life (principally cans and bottles) need to be stabilised

before packaging. The most common (and troublesome) form of instability is the production of a non-biological haze, or cast, in the product. Non-biological haze material will be precipitated, along with yeast, during a low temperature storage period. There are two components to non-biological haze, both of which are capable of producing particles large enough to produce a visible cast to the beer. Hazes are mainly produced by interactions between low-molecular weight proteins, tannins and carbohydrates. Their two forms are:

(a) a chill haze, which involves weak electrostatic bonding between the constituent molecules. Such a haze is produced when the temperature is lowered, but disappears when it is raised again, *i.e.* these hazes are reversible.
(b) a permanent haze. These are the result of permanent chemical bonds between (usually) proteins and polyphenols (tannins) and are unaffected by temperature. Production of such bonds is often catalysed by the presence of heavy metal ions (especially Sn and Pb) and is promoted by the presence of oxygen.

As well as removing the haze once it is formed (by precipitation), it is also possible to remove it by precipitating one of the haze precursors. Proteins can be removed by the addition of an insoluble stabiliser, such as bentonite or silica hydrogel (lucilite), which will precipitate the protein quite rapidly in the conditioning tank. Tannic acid can also be used to bring down proteins, but is little used because of the copious volume of highly viscous bottoms (sediment) produced. Polyphenols can be removed by adsorption onto polyvinylpolypyrrolidone (PVPP, or polyclar).

Hazes can also be broken down enzymatically into smaller particles not visible to the naked eye. Papain is the most widely used of such enzymes, but it has to be suitably ingredient-labelled because it will end up in the finished product; this situation does not apply to insoluble stabilisers because they are removed from the beer stream once they have performed their function.

Other contributors to non-biological haze, such as β-glucans, can also be removed enzymatically if necessary.

It is now generally accepted that the use of high quality raw materials and good brewing practice can largely overcome haze problems.

(4) *To carbonate the beer.* To some extent, carbonation occurs naturally during the cold storage maturation period, but only relatively small volumes of CO_2 are liberated. The gas will, of course, be more soluble at reduced temperatures. It is common practice, therefore, to maintain a top pressure of CO_2 in a beer conditioning tank.

(5) *To minimise dO_2 in the beer.* At the termination of primary fermentation, green beer should have a low dissolved oxygen (dO_2) content. Transfer of beer during post-fermentation brewery operations provides the most likely means of imparting oxygen once again (*e.g.* by faulty pump seals). There is even more chance of increasing beer dO_2 once the yeast has been removed. Antioxidants can be added, the commonest being potassium metabisulfite, sulfur dioxide and ascorbic acid. SO_2 has to be used with care since there is a legal limit to its concentration in beer ($\leqslant 70$ ppm).

Because there are many different conditioning and maturation regimes available to the brewer, so there are characteristic maturation profiles – these indicate the relationship between temperature and the amount of yeast in suspension. Thus, an ale maturation profile is different to a classic lagering profile.

There are two vastly different ways of beer conditioning and these will produce greatly differing end products:

(1) cask-conditioning (draught beer);
(2) brewery-conditioning (keg, bottled and canned beer).

CASK-CONDITIONED BEER

Draught beer is a living product; the 'real ale' of CAMRA. Yeast is left in contact with the beer in the container (cask) in which it is placed (racked) and this allows secondary fermentation to take place. During its time in the cask the beer will undergo flavour maturation, carbonation (naturally produced during secondary fermentation), reduction of dO_2 and stabilisation. A clarification (fining) agent can be added immediately, or just prior to the cask being dispensed into trade. The beer can be racked into casks, either directly from the fermenter, in which case very little yeast will have separated out, or it can be passed into a racking machine (normally an elongated tank with multiple outlets leading to racking lines and valves). If necessary, some yeast settlement can occur in these tanks (often called racking backs) and this enables the operator to maintain a constant volume of yeast cells per ml of beer. Too few yeast cells will cause under-conditioning in the cask (and hence flat and uninteresting beer), whilst too many cells can cause over-conditioning, with concomitant dispense problems in trade. Depending upon the brewery regime and the type of beer to be produced, the concentration of viable yeast cells in racked cask beer can vary from 3×10^6 to 2.5×10^5 cells per ml.

One of the major problems encountered during cask-filling is fobbing (or excess frothing), but this can be eliminated by using a modern counter-pressure racking device.

Casks

For an illustration of the main parts of a wooden cask see Figure 6.1. Casks were originally made of oak segments held together by iron bands, called hoops. The preferred wood used to originate from Russia and Poland (for reasons of workability and ability to store beer without imparting any unwanted flavours). The main body of the cask consists of curved planks called staves, one of which is wider than the rest (the bung stave) and holds the bung hole. The ends of the cask, called top head and bottom head, also consist of individual wooden sections (usually four per head) called middles and cants. The latter

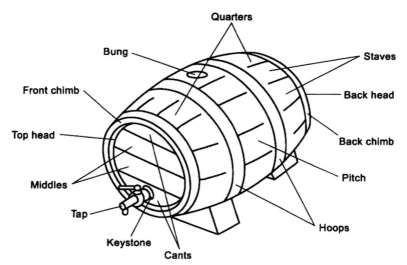

Figure 6.1   *Parts of a wooden cask*
(Courtesy of the Campaign for Real Ale, St. Albans, Herts., UK)

are D-shaped and one of them on the top head will hold the keystone. The heads are held in place by the overlapping ends of the staves (surrounded by a hoop) and these are referred to as chimbs (front and back). Both the bung hole and the keystone are seated into the cask by means of metal bushes, normally constructed of brass. The bung hole is sealed by a bung, or shive, which has a thinner inner section (called the tut), whilst the keystone, which holds the tap when the cask is in use, is initially sealed with a faucet plug which also has a reduced central section for receiving the tap. Shives and faucet plugs were originally constructed of wood, but plastic versions are commonly used these days.

Wooden casks have to be 'seasoned' before use, usually by filling with boiling liquor containing soda or salt, storing for 24 hours and then emptying and steaming. If not being used regularly, surplus casks are often stored with the top head removed (to allow better aeration) and the inner surfaces brushed with bisulfite of lime. Surplus wooden casks often developed a mustiness after a period of prolonged inactivity and these would be 'freshened' by igniting sulfur matches inside them.

Wooden casks are notorious for their porosity, not only by way of the joins between staves, *etc.*, but by the very nature of the wood itself. It was common practice to line the inner surface of casks with 'brewers pitch'. There were various types of pitch, all basically consisting of heated (to drive off volatiles) resins from coniferous trees diluted down in linseed oil, paraffin wax, or something similar. This mixture would then be painted or sprayed whilst still hot onto the inner surface. On drying, a film would be produced which prevented excess leakage. This lining would, of course, soften in warm weather! Even enamelling has been tried, the enamel being sprayed on, tri-layered, in a solvent (such as trichloroethylene). With the robust treatment afforded to casks, the enamel layer was very prone to fissure. Wooden casks are still used by certain UK breweries (*e.g.* Wadworth & Co., of Devizes, and T. & R. Theakston Ltd, of Masham).

Some of the nomenclature applied to wooden casks is pertinent to their metal successors, although the latter are constructed by welding pre-formed castings together.

Most casks these days are constructed of aluminium or stainless steel, although trials are being carried out with polymer containers.

Aluminium has several advantages over other materials, including its low cost in relation to its strength, together with its ease of working. It also shows high thermal conductivity. It does, however, have a tendency to corrode (especially in the presence of acids and alkalis) and it is not really compatible with beer. Corrosion was shown to be enhanced when kegs with stainless steel spears were first introduced. When filled with beer, a galvanising effect was produced, with the beer being used as an electrolyte in a 'battery-like' situation. Thus, aluminium has to be protected on the inner surface of the cask; the original means of protection was by anodising. This is, in effect, the controlled oxidation of the metal, using sulfuric acid and electricity. The anodic oxide film grows in a needle-like fashion and then has to be sealed with superheated steam which causes it to swell. The sealed film is usually some 10–15 μm thick, completely inert and corrosion-resistant. Unfortunately, the film proved to be very hard and brittle and prone to fissure when casks were handled heavily.

Anodising fell out of favour in 1985 when alochrome became more widely used. Alochrome is produced by a process which uses a mixture of chromic, sulfuric, hydrofluoric and phosphoric acids. The solution is carefully introduced onto the inner surface of the cask and then heated at a specific temperature for a precise period of time. This converts the inner surface to a very complex aluminium–chromium–sulfate–fluoride film which is inert but not brittle. When cool, an epoxy–phenolic lacquer is applied before the cask is stoved. The resultant film is some 25 μm thick, completely inert and extremely long-lasting. The only way to disrupt it is to damage it physically by abrasion.

Stainless steel is now the most widely used material for beer containers. Stainless steels are a group of steels containing no less than 9% chromium, and many contain other alloying elements. Their high corrosion-resistance is conferred by a naturally-occurring chromium oxide layer. There are three main ranges of stainless steel: austenitic, ferritic and martensitic, of which only the former is used in the brewing industry. Within the austenitic range only grade 304 is used in cask manufacture, a typical analysis of which is:

Element	% composition
C	between 0.03 and 0.07 max.
Si	1.0 max.
Mn	2.0 max.
P	0.045 max.
S	0.030 max.
Cr	17.0 min. to 19.0 max.

Table 6.1 *Cask sizes*

Description	UK gallons	US gallons	Litres[a]
Hogshead	54	64.5	246
Barrel	36	43	164
Kilderkin	18	22	82
Firkin	9	11	41
Pin	4.5	5.5	21

[a] 4.6 litres per UK gal., 3.8 litres per US gal.

Cask Sizes

Even as we enter the next millennium, traditional beer in the UK is still largely sold in non-metricated containers, the principal cask sizes being illustrated in Table 6.1. Nowadays, most beer is conveyed in either kilderkins or firkins.

Cask Cleaning

One of the on-going problems in the brewing industry is the subject of cask cleaning. There is absolutely no point introducing sound beer into microbiologically unsound containers.

Originally, wooden casks were de-headed (top head) and scrubbed internally, a laborious process which eventually gave way to a process called 'running in'. This involved filling swilled casks with boiling water, sealing them and leaving them overnight. The casks would then be emptied, rinsed and steamed. Both processes inevitably cause gradual damage to the inner surface of the cask.

Another early cleaning method involved the 'rumbler'. In this method, one gallon of water and some clean gravel were added to a cask, which was then sealed, placed on a pair of rollers and rotated for a period of time. This method scours (and damages!) the inside of the cask and was only used for casks with very heavy internal deposits.

Pitched casks could not be subjected to hot water or to the rumbler.

Wooden casks have always been notoriously difficult to clean, especially when it is realised that certain bacteria and fungi can penetrate up to 6–7 mm into the wooden fabric, and that boiling water and/or steam can only surface sterilise unless carried out for prolonged (impractical) periods.

The first attempts to automate the washing process yielded some variable results, in terms of internal cleanliness of casks, a fact that was highlighted when simple bacteriological techniques became more widely used. Modern machines, with their ability to inject high pressure steam, *etc.*, are now highly efficient.

Each brewery will have its own specifications and requirements, but a washing cycle normally involves:

—an external wash;
—bung and shive locating and removal;
—a rinse to remove any ullage (a 'gunge rinse');
—a hot water rinse;
—a hot detergent rinse (probably two) at 60–70 °C;
—a hot water rinse (probably two);
—a steam sterilisation;
—an inspection internally (visual and microbiological).

The Gimson 'Cask-Major' (Figure 6.2) is a well-tried machine and works on the walking beam principle whereby casks are collected, lifted, transferred to a station and then lowered. The cycle is as follows:

—Station 1. Cask is received onto the machine.
—Station 2. Cask is given external wash; the bungs are located and cask is subjected to an internal wash. The water used has been reclaimed from Stations 4 and 5, and is fed at 100 psi and at 20 gpm (gallons per minute).
—Station 3. Internal detergent wash. Detergent fed at 100 psi at 10 gpm.
—Station 4. Internal hot water wash. Water fed at 100 psi at 20 gpm.
—Station 5. Internal hot water wash followed by steam treatment. Water is fed at 100 psi at 20 gpm. Steam is introduced at 40 psi and at a rate of 175 lb h^{-1}.
—Station 6. Cask is discharged for inspection.

The temperature of the detergent at Station 3 must be at least 70 °C. The hot water for internal rinsing at Stations 4 and 5 should be at least 90 °C. Contact time for the critical phases (Stations 3, 4 and 5) should be at least 60 seconds. Adequate drainage time should be allowed between Stations 2, 3, 4 and 5. Flooded casks do not clean properly.

Casks which fail to respond to a conventional washing cycle are called 'stinkers'. At Nethergate, where we only use stainless steel casks, the stinkers are soaked in warm caustic soda overnight.

The pricipal problem with casks, in terms of cleaning difficulties, is

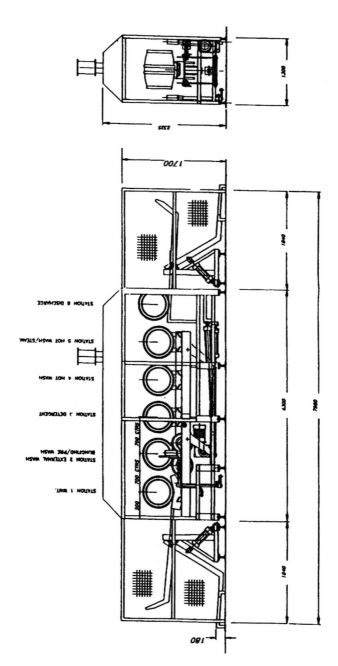

Figure 6.2 *Diagram of an automatic six-station cask washer*
(Courtesy of Gimson Engineering, Beaumont Leys, Leicester, UK)

the build-up of beer-stone (which is basically a build-up of calcium oxalate), protein deposits and yeast deposits. Beer-stone generally has to be removed by acidic (where possible) or highly sequestered detergents.

After a cleaning regime, casks must always be cooled before filling with beer or being checked for microbiological sterility.

Fining

To permit draught beer to drop bright when finally in trade, finings are added; either directly to the cask or to the racking back itself. Beer produced by earlier generations was undoubtedly cloudy when drunk, even though the product would have been stored for some time in bulk vats. (Storage itself will facilitate some degree of clarification by pure sedimentation.) Beer during these times, however, was not imbibed from glass vessels and so its appearance mattered little; it would certainly have been consumed cloudy up until the end of the 16th century.

When commercial brewing commenced during the latter part of the 18th century, factors such as clarity became more important, for wine as well as beer. All sorts of things were used to effect clarification, including chalk, Fuller's Earth, wheat flour, bean flour and oyster shells – and often accompanied by a touch of treacle! Before the advent of sound earthenware vessels, beer and wine used to be stored in dried animal skins and other containers of animal origin such as stomachs and air sacs (swim-bladders of large fish). The Romans noticed that wine stored in animal stomachs and air sacs was much less opaque than that stored in any other kind of container.

Over the years many substances have been employed to clarify beer, including egg-white (albumen) and gelatin, but the most efficient has proved to be isinglass. Isinglass is the dried swim-bladder from certain species of fish; the word is a corruption of the German 'hasenblase', which means 'hare's bladder'.

In Europe, isinglass was first used commercially in the late 18th century, and the major source was the sturgeon (*Acipenser* spp.) which was found in many large European rivers, such as the Rhine, Volga and Danube. In the UK, the Trent and the Severn were known sources of this fish, which produces very high-quality finings. Nowadays, the main sources of isinglass are from fish caught in waters $10°$ either side of the equator; such species as catfish, drumfish and threadfins. The shape of the swim-bladder (and the quality of the final product) exhibits much variation and gives rise to exotic names such as Round Saigon, Long Saigon and Maracaibo Pipe. To prepare finings fit for brewery use, the isinglass is removed from the fish and dried naturally (*i.e.* hanging out at ambient temperature). If dried too rapidly the product will lose much of its clarification potential. The dried 'leaf' is then cleaned, sterilised in hydrogen peroxide and 'cut' in acid. Phosphoric acid is used now, but hitherto, tartaric, citric and sulfurous acids have been used as cutting agents. Cutting produces a viscous white liquid which can be stored and transported. Original production methods employed whole leaf and the process of produc-ing usable finings took six weeks. Modern technology allows the leaf to be rolled and shredded and finings to be produced within two days. Some types of leaf are easier to process than others and these are more highly regarded and, hence, expensive. Such leaves also happen to produce the best liquid finings (*e.g.* Round Saigon).

The fining action of isinglass is attributable to the protein collagen which is present in large quantities. On acid hydrolysis, isinglass releases individual triple polypeptide helices of collagen, each contain-ing numerous positively-charged sites. The mechanism of fining is thought to be associated with electrochemical interactions between these sites on the collagen molecule and the net negative charge on the yeast cell surface, thus promoting the sedimentation of large numbers of cells.

Although the principal component of isinglass finings is collagen, small amounts of gelatin are also present. Gelatin is a degradation product of collagen, and in some parts of the world, notably the USA, it represents the main means of fining beer. The sources of gelatin are normally animal bones and hides. Gelatin has a clarification power approximately one-twentieth of that of collagen.

Collagen can also be obtained from bovine sources, although, with the concern over BSE in cattle, this particular market is now some-what limited. Bovine collagen gives an inferior clarification perfor-mance when compared with that obtained from fish, and this is probably due to the de-stabilising of the protein during the 'de-hairing' process of the animal which involves alkali treatment.

Recently, James Vickers in the UK have developed and patented a microbial de-hairing method which employs fermenting cultures of *Lactobacillus plantarum*. The bovine collagen produced by this method has much greater clarification powers than other samples.

The thermal stability of collagen from different sources exhibits great variability and this has been attributed to the amount of cross-linking between the individual helices; the higher the degree of cross-linking, the greater the thermal stability. Collagen from certain sources can even denature at about 15 °C, rendering it totally impractical for brewery use. All isinglass finings lose their integrity rapidly at temperatures above 25 °C, and storage in the brewery, in liquid form, should be between 4 and 10 °C.

Isinglass finings will also react with lipids and any negatively-charged proteins within the beer and these will precipitate out with the yeast.

Under certain circumstances, yeast cells exhibit only a weak negative charge and this can lead to clarification problems. To overcome the possibility of the occurrence, auxiliary finings are usually added to the beer prior to isinglass. These are normally silicate- or alginate-based and are strongly negatively-charged so that they enhance the yeast surface charge. Underfining and overfining with isinglass can also lead to poor clarity in beer and it is important to ensure that the exact amount is added. This is a matter, initially, of trial and error, and every brewery will adopt a slightly different regime. As a rule of thumb, the normal rate of prepared isinglass addition is around two pints per barrel. Once introduced into a cask with beer, finings gradually deteriorate. This is exacerbated if the beer is moved in transit too many times (*i.e.* finings only 'work' for a limited number of times).

Dispensing of Cask-conditioned Beer

For centuries, when brewing was confined to religious establishments, grand houses, farms and the like, ale would have been dispensed by inserting a tap into the cask and drawing it off by gravity. This method would also have been used by the numerous victualler brewers (*i.e.* pub-breweries, where ale was brewed and sold on the same premises) which predominated in the 17th and 18th centuries (there were some 48 000 recorded in the UK in 1750!). Beer in its casks would either have been stillaged (laid out horizontally) on a gantry behind the serving area, or kept in an underground cellar in large containers (butts) from where it would be brought upstairs in jugs by 'pot-boys'. The latter method was time-consuming and very labour-intensive, especially in a busy inn; nevertheless, the publican could make a good living. When 'common brewers' (*i.e.* commercial brewers) began to establish themselves in the 19th century there was a greater emphasis on hygiene and general serving protocol; such brewers wanted their beer to be presented in the best possible condition. Competition was immense and as brewers' profits rose so beer prices increased. This put pressure on the publican, who was buying the beer in, and led him to search for faster and less labour-intensive methods of dispensing beer. A method of mechanically drawing beer up from the cellar would be very useful.

The first recorded invention was by Joseph Bramah who, in 1797, patented the first beer engine. The system involved the use of 'sand pressure', whereby heavy boxes of sand, with an associated pulley-system, would operate a crude piston which forced beer up to the bar from an underground container. Unfortunately, beer had to be transferred manually from the butt supplied by the brewery to the container capable of allowing the beer to be shifted upwards. The apparatus proved to be nearly as labour-intensive as bringing beer up from the cellar in jugs and it necessitated cellars of considerable height to accommodate the sand boxes and pistons. The pipework involved would, almost certainly, have been made of lead! Bramah's idea did not meet with any general acceptance, but it did stimulate others to try to find an answer to the problem, and by 1801, 'beer-engine makers' were registering themselves in London (*e.g.* John Chadwell of Black-friars Road). Similar devices, used for moving water, had been in production for several years previous to that and were especially common in maritime communities, but no-one had seen fit to uplift beer with them. Once the natural antipathy had been overcome ('beer should be drawn by tap straight from the wood') their use spread rapidly, and by the end of that decade suction beer pumps were very common in London and other large towns. By the 1820s, beer service by engine had become a standard feature of inns throughout most of urban Britain. Some of the large London taverns were reported to have had banks of 40–50 handpumps. The earliest pumps would have been constructed with lead pipework and a leather piston – the 'leather bucket' engines. Lead gave way to porcelain and/or glass (segments held together with rubber joints), and leather yielded to all-metal piston manufacture, usually of gun metal or brass. The modern handpump has its working parts (especially those coming into contact with beer) constructed of stainless steel and plastic, whilst the pipe-work consists of food-grade micro-bore plastic.

The handpump, in its modern form, is a simple lifting pump consisting of a pulling handle, which, *via* connecting rods and a pivot, operates a piston contained in a cylinder. The handle itself is firmly attached to the bar counter; the rest of the apparatus is mounted underneath (and normally invisible from the customer side of the bar). The base of the piston and the base of the cylinder contain non-return valves to ensure one-way flow of beer from inlet pipe to outlet (dispense) point. Cylinders may be mounted in a horizontal or vertical fashion.

When the pump is not being operated the handle is normally in a vertical position on the bar. Both piston and cylinder valves will be closed.

To draw up beer, the handle is pulled downwards causing the piston to rise. This, in turn, creates a suction pressure which causes the valve at the base of the cylinder to open. Beer is then drawn into the space below the piston. The pressure above the piston (outlet side) is greater than that below (inlet side) and so the piston valve remains closed.

The handle is now physically returned to the upright position which causes the pressure underneath the piston to exceed that above it. Thus, the piston valve opens and the piston passes down through the beer contained in the cylinder. The positive pressure differential below the piston will have caused the lower cylinder valve to close.

The handle is now pulled down again and the beer enclosed in the cylinder (which is above the piston) is drawn up by the piston and escapes through the outlet pipe. The handle is then returned to its upright position and the cycle continues whilst serving is in progress. The volume of the cylinder (*i.e.* one pull) is normally one half pint.

If beer lines are left empty between serving sessions (*e.g.* overnight) then the first two or three 'pulls' of the new session will effectively be priming the pump.

Hand-pulled beer engines are most effective when the bar counter and serving area are immediately over the cellar, thus giving a fairly short pipe run between cask and pump. This situation does not always arise, however, especially in the more modern pubs and clubs, and some assistance may be necessary to draw up beer.

Electric metered pumps were commonly employed for this purpose and still represent the main means of dispense in certain areas of the UK. The electric pump, which is normally situated in the cellar (on the wall!), operates a diaphragm in a glass chamber (situated under the counter) which discharges the contents of the chamber when the switch (which is counter-mounted) is turned on. Very accurate measures can be dispensed, the exact volume being dependent on the size of the chamber.

Because of modern safety regulations, electric pumps have tended to be replaced by gas-powered pumps; nitrogen, carbon dioxide or compressed air are used to aid dispense. The gas does not come into contact with the beer.

In some outlets the beer is necessarily stillaged at a level above that of the point of dispense (handpump). To prevent slow leakage through the pump a check valve must be inserted between the cask and the handpump.

After racking into casks in the brewery, beer should be stored in a cellar, or other refrigerated area, to allow some secondary fermentation (conditioning) to take place. From store it is transported to the retail outlet by dray, where it is then placed into the cellar and set up. Assuming that a cellar or refrigerated room is available, there are two main ways of preparing a cask for subsequent dispense: they can be laid horizontally (stillaged) or set up vertically.

When stillaging, one can use the floor or a permanent rack (or gantry). Unless the casks have just been delivered, they should be rolled around the cellar floor before being set in position. This ensures thorough mixing of finings. Casks are then chocked into position using three wedges of wood: two at the front and one at the back. The cask is always chocked on the hypotenuse of the wedge, and it should be ensured that the cask itself is supported only on the chocks. The cask should be tilted slightly, with the back head higher than the front head. This ensures that the sediment will collect and settle in the belly of the cask. If the cask is tilted too steeply then the sediment will collect around the keystone which will lead to clarity problems and high wastage levels.

The cask can now be vented. This is brought about by inserting a wooden peg (spile) into the recess (tut) of the shive. A softwood (bamboo) spile should be used initially and, once inserted by means of a mallet, pressure within the cask (caused by slow secondary fermentation) is released. Results can be spectacular, depending upon the amount of condition in the beer. For beers that are notoriously lively, venting is often carried out with a 'controlled vent peg'. This is a metal device with a needle valve, to permit gas release, and a side arm through which ale can escape for collection in a clean container. Such fobbed beer can be returned to the cask (as soon as possible) using a tun dish. The fine perforations along the stem of the funnel allow beer to be replaced without disturbing the contents of the cask. This is the only time that it is permissible (advisable) to return beer to a cask!!

The cask is now ready to be tapped. Various theories exist as to when tapping should occur, but for most beers the tap can be inserted almost immediately after pegging. The tap is driven through the recess in the keystone with a mallet; a single blow should suffice. Some landlords tap the cask (placed in an upright position) before putting it on the stillage and spiling it, thus using the tap as an initial venting tool.

Once effervescence through the soft spile has ceased, it is usual to replace it with a hardwood spile, which effectively seals the container again.

Twenty-four hours are allowed (time can vary) for the sediment to settle out and then the beer is checked for clarity by withdrawing a sample.

If cellar space is limited and horizontal stillaging is impractical, then casks may be stood in an upright position (on their back chimbs) and an ale extractor (siphon) used to withdraw beer. With this method the shive is not broached in any way. This is called vertical extraction. Venting is performed by driving a shank, with a blanking plug in it, into the keystone. The shank possesses a lateral venting cock, which should be in the closed position when the shank is being driven in. Once the shank is inserted the cock is opened to release the pressure. Excess fob can be collected if necessary. Once effervescence has ceased, the blanking plug is removed and the extractor tube inserted. The tube is placed to the very bottom of the cask and then raised approximately one inch before locking it in position. The venting cock is then closed until the cask is required for use. To avoid sediment collecting around the base of the cask beneath the keystone, it is usual to place a small wedge underneath the cask to encourage sedimentation to occur away from the keystone side.

Ale in casks set up for vertical extraction cannot easily be assessed for clarity before serving, *i.e.* there is no tap from which to withdraw a sample. One can, however, siphon out a sample through the keystone aperture. One slight advantage of this method of dispense is that it is possible to use the top (settled) layer of beer by simply inserting the extractor shaft into that region. One can then 'follow down' the cask by gradually lowering the shaft. The dispense end of the extractor shaft will be attached to the beer line by a clip.

For horizontally-stillaged casks that are to be dispensed by hand-pump, the tap will terminate in a thread (male) which receives a compatible union (female). The union, or burr, surrounds a tail which is connected to the beer line. When connecting the union to the tap, a hop filter can be inserted between the two to prevent particulate matter (*i.e.* from dry-hopping) from entering the line. If a filter is not required then a washer should be used.

The ideal cellar temperature for traditional ale is 12–14 °C (54–57 °F). Some very old underground cellars are capable of maintaining this, but most need some kind of cooling equipment. When installing cellar cooling one should take into account the following:

—the size of the cellar;
—the thermal conductivity of the walls, ceiling and floor;
—the volume of beer to be stored.

If placed in the cellar, the chilling unit should be positioned such that it receives a good air flow, but that it does not emit warm air in the vicinity of the casks. If practical, warm air should be emitted through an external vent. Where possible, it is advantageous to site chillers outside of the cellar. Although beer may be at the desired temperature in the cellar, there is always a danger that the length of the dispense line (and the ambient temperature in which it is situated) may cause it to rise by the time the dispense point is reached. Lagging the lines can help, but these days a 'python' system is often used, especially in the larger ale-houses. These maintain beer temperature from cellar to bar. A python consists of multiple beer lines held together in an insulating butyl rubber surround, the whole arrangement being encased in a tough plastic skin. Through the centre of the python run two chilled-water recirculating lines, ultimately connected to a chilling unit. Lager (or keg beer) lines immediately surround the chilling lines, since these products are dispensed at a lower temperatures than cask ale. The ale lines would be around the periphery of the python.

Tubing used for beer dispense must be constructed of materials that conform to International Statutory Food Regulations. Nowadays, one of the most widely used compounds is natural polyethylene, also known as medium density polythene (MDP). The material is tough, flexible and opaline. The plasticisers are of a very specific type and are used in very low concentrations.

Some of the earlier 'plastics', such as polyvinyl chloride (PVC), have been found to be unsatisfactory for use about foods because of toxicity from plasticisers.

Nylon can be used, but it is much more expensive and so its use is confined to the production of narrow-bore, high-pressure dispense lines (*i.e.* keg beer systems). It exists in two forms: semi-rigid and flexible. Only the former is used in the brewing industry.

Some landlords still dispense beer by gravity, the casks normally being stillaged behind the bar. This looks very appealing to the customer, but it has its drawbacks, not the least of which is temperature control in the warmer months. The simplest way to keep casks cool is to place wet sacking over them and moisten them frequently. This is called evaporative cooling. A far more efficient way to cool such casks is to use an in-cask cooler. This consists of a cooling probe which is inserted into the bung of the cask (where the shive would normally be). The probe is connected to a chiller and cold water is circulated through it. The apparatus is quite efficient, but, since the probe is actually situated in the beer, it must be meticulously clean before it is used.

To enable as much beer as possible to be drawn from a horizon-tally-stillaged cask, it must be tilted forwards carefully at some stage of its life. This is simply carried out by removing the rear chock; raising the back chimb until the top is level with the shive and re-inserting the chock (being careful to avoid settling of sediment around the keystone). This should never be attempted if the cask is less than one-third full (it is ideally done when it is about half full). If possible, tilt at the end of a serving session so that, if there has been a slight disturbance of the beer, it will have time to re-settle.

Over the last few years a number of improvements to the means of draught beer dispense have been introduced. These have included the self-tilting stillage and water-cooled handpump cylinders, as well as the siphon itself. A glance at Samuel Mason's (now Harry Mason) catalogue of June, 1903 (Figure 6.3) shows us that very little has changed over the last hundred years or so!

When beer is drawn from the cask, by whatever means, the spile will have been removed so that a free-flow can be effected. This means that air will replace the beer that has been removed. Air (particularly oxygen) is the worst enemy of cask beer (apart from an uncaring landlord!) and contact between air and beer should be kept to a minimum. Replacing the hard peg after every serving session helps enormously, but some oxidation is inevitable. Ideally, a cask should be consumed within 3–4 days, regardless of its size. For situations where turnover on a product is slow it is permissible to use a 'cask-breather' to prevent oxidation. The cask-breather is, in effect, a demand valve which allows carbon dioxide to enter the cask as beer is being drawn off.

The inlet to the device is connected to a source of CO_2 *via* a regulator. The outlet is connected to the spile hole in the shive, either directly or *via* tubing.

As one pint (say) is withdrawn from the cask, the valve allows 'one pint' of CO_2 into the cask to replace it. This occurs at atmospheric pressure, there being no change in the gas content of the beer; the dissolved CO_2 balance thus remains unaltered.

A second, one-way valve in the breather allows excess CO_2 to escape should secondary fermentation be vigorous (*i.e.* it can act as a soft spile).

As well as excluding air, airborne micro-organisms are also denied entry to the cask and so the shelf-life of the product can be consider-ably prolonged. If airborne contamination (either by microbes or some natural odour) is suspected then an air filter can be inserted, *via* tubing, into the spile hole.

No. 1833   Cask tilter   12/6 each

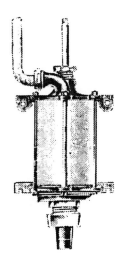

Stone cylinder
Always clean and cool . . . . 21/6 each

No. 1813   Extractor for
barrels with single connection
10/6 each

Figure 6.3   *'Nothing's new!' Illustrations from Samuel Mason's catalogue of
June, 1903*
(Courtesy of Harry Mason Ltd, Sun Works, Birmingham, UK)

Carbon Dioxide

The dissolved CO_2 content of a cask beer is an all-important entity and its presence in too great or too small an amount will affect the palatability of the product. Carbon dioxide is very soluble in beer, only marginally less so than it is in water. Variation in its solubility in various beer styles is marginal.

If gaseous CO_2 is in contact with beer in a closed container (*e.g.* a cask) then a transfer will occur between the gas/liquid interface until a dynamic equilibrium is reached. This is maintained as long as the temperature and pressure remain constant. What is happening, at this point, is that the rate of transfer of CO_2 from the liquid to the gaseous phase is being counteracted by CO_2 transfer from the gaseous to the liquid phase. Therefore, for any given temperature and pressure, a beer will not keep on absorbing CO_2 merely because it is in contact with it.

According to Henry's Law [equation (6.1)], the concentration of CO_2 in a beer, held at constant temperature, is directly proportional to the partial pressure of the gas (P_{CO_2}) in the gaseous phase,

$$P_{CO_2} = Hx_{CO_2} \qquad (6.1)$$

where x_{CO_2} is the molar fraction of CO_2 in the liquid phase, and H is Henry's constant.

If the temperature changes, so does Henry's constant. Thus, at a set pressure, if the temperature of the beer decreases then the solubility of CO_2 increases.

The amount of dissolved CO_2 in a beer is normally expressed in terms of 'volumes per volume', and under normal cellar pressure and temperature conditions, a pint of beer contains slightly over one pint of dissolved CO_2.

A sound, unbroached cask, as delivered to an outlet, will contain CO_2 as the sole gas, and the pressure inside the cask is above atmospheric until the soft spile is introduced. This induces a release of undissolved gas. As the pressure reverts to atmospheric, more CO_2 will gradually be released from the beer and will escape from the vent peg until the dynamic equilibrium is re-established. By definition, at 15 °C and 1 bar partial pressure of CO_2, there will be 1.1 volumes of CO_2 dissolved in one volume of beer.

When beer is drawn off, and air enters the cask, the situation changes because P_{CO_2} decreases as its percentage of the total gas mixture decreases. This causes more CO_2 to be slowly released from the beer in an attempt to re-stabilise the situation. As a result, the beer will start to 'lose its condition'. Remember, the whole container is still at atmospheric pressure (and, hopefully, constant temperature).

Since the loss of CO_2 from beer is a relatively slow process, it can be seen that if a cask is used within a period of two or three days then the beer being dispensed from it will always have reasonable condition. Keeping the hard peg down during non-serving sessions helps maintain condition in the cask.

The importance of a constant, cool cellar (and beer) temperature has already been explained, particularly with respect to its relationship to dissolved CO_2. Extremes of temperature are hazardous as well. At low temperatures the beer can throw a 'chill haze', which is normally caused by proteins being precipitated. The haze can be irreversible and, whilst it will not completely ruin the flavour of the beer, it will certainly make it visually unattractive to drink.

At the other end of the spectrum, temperatures around 23 °C (74 °F) and above can cause denaturing of finings.

Cellarmanship, like brewing itself, is a mixture of art and science, and for a loving account of the subject the reader is recommended to Ivor Clissold's book.[1]

BREWERY-CONDITIONED BEER

Brewery conditioned (or processed) beer is that which has been conditioned, chilled, filtered and pasteurised in the brewery before being introduced into sterile containers such as kegs, cans or bottles. Thus, the end-product is technically 'dead', since all viable microorganisms, including yeast, have been removed.

Bottled beers have been documented since 1568 and in the early days beer was introduced into bottles by decanting straight from the cask; no treatment was used at all. Corked bottles (or Hamilton bottles) were introduced in 1814, but the major innovation was the development of the metal crown-cork by William Painter in 1891. The crown-cork enabled modern bottling technology to be developed, culminating in the highly sophisticated, automated machines of today. At the turn of the century, vast quantities of beer were bottled and transported around the country, one of the most successful being Whitbread & Co. The Chiswell Street brewery commenced bottling in 1870, producing some 1293 barrels per annum in this form. By 1912 the brewery was bottling 45% of its total output (439 532 barrels), according to Redman.[2] Cans and kegs, on the other hand, are relatively recent introductions. The first canning line was installed by the Felinfoel Brewery, Llanelli in 1936, whilst kegging, which was first introduced by J.W. Green of Luton in 1946, did not really become the vogue until the 1960s when lager beers became popular in the UK. It is estimated that some 60% of all draught beer produced in the UK today is in keg form.

The first stage of the brewery-conditioning process is to introduce green beer into a conditioning tank, where maturation, stabilisation, *etc.* can take place. If a Unitank has been used for primary fermentation, then green beer transfer is unnecessary. Conditioning tanks are closed vessels, normally constructed of stainless steel and take two main forms: horizontal and vertical (cylindroconical). The latter are now favoured, mainly because of reduced floor space required, ease of cleaning, reduced beer losses and reduced beer surface area for oxygen pick-up. Vertical tanks are usually of greater capacity than horizontal tanks (which rarely exceed 500 barrels). The advantages of horizontal tanks are a more rapid sedimentation, ease of mixing and better temperature control.

Both designs of tank are equipped with cooling jackets (these can be difficult to position correctly on cylindroconicals), a single inlet/outlet, temperature probes and a cleaning in place (CIP) system.

Exactly what happens in the conditioning tank now depends on the nature of the final product and its intended shelf-life once in trade. High-quality, bottom-fermented, continental lager beers traditionally undergo secondary fermentation over a period of weeks, or even months (with the temperature gradually being reduced from 10 °C to 0 °C), followed by a similar extended period of cold storage. This facilitates the precipitation of yeast, cold-break material and chill haze material. Beer leaving such vessels will be almost 'bright' before being introduced to filters prior to further processing. Flavour maturation will also occur, slowly, during cold-conditioning as long as there is still some yeast in suspension.

Because 'time equals money' there has been a trend towards reducing lagering times in breweries to periods of two or three weeks (or even a few days in the case of some mass-produced products!).

Some aspects of flavour maturation (notably the removal of diacetyl) can be expedited by elevating the beer temperature for a while. For example, beer held for a few days at 12–16 °C will show rapid removal of diacetyl. This operation is referred to as ruh storage (or the 'diacetyl rest'). After this warm-conditioning, the beer is then subjected to a cold-conditioning stage (up to ten days at 0 °C) before any further processing.

In some breweries, primary fermentation proceeds to a level

whereby there is very little fermentable material left for secondary fermentation in tank (*i.e.* less than 1% of the original gravity). It is, therefore, necessary to add fermentables to allow the yeast to undergo its various purging activities. Sugars (invert sugar) or fresh wort can be added, but one of the traditional, continental ways of achieving this is to add a batch (about 10% of the total volume) of freshly fermenting (about 3 day-old) wort to the conditioning tank. This is called krausening and comes from the word 'krausen' (meaning a frill) which describes the appearance of the yeast head typical of bottom fermentations of this age. In the UK these are called cauliflower heads.

The Finnish brewers Sinebrychoff have developed a system whereby maturation of beer can be reduced to about two hours! An immobilised yeast system is employed. The carrier for the yeast is Spezyme GDC (manufactured by Cultor, Finland) which has advantages over other carriers such as calcium alginate. Spezyme GDC is the carrier used for the glucose isomerase enzyme in the production of high-fructose corn syrup; it is, therefore, approved for food use.

The carrier consists of cellulose, as a diethylaminoethyl (DEAE) derivative, which is the active binding site for the yeast cells; and polystyrene, which is the support for the cellulose and titanium dioxide, and also adds weight. The unit in which the active ingredients are contained is called the reactor. The efficiency of the process is due to the vast number of yeast cells and an extremely good contact between yeast and beer.

To start the maturation, green beer is first heated to 90 °C for seven minutes to convert diacetyl precursors to free diacetyl. It is then pumped through the reactor; no chemicals are used. Once initiated, the process can be run continuously, normally for a period of about six months before the yeast has to be replenished. If necessary, the reactor can be closed down for a short period (*e.g.* a weekend) without any deleterious effect on the yeast, start-up taking about one hour. If the plant is shut down for longer periods then the yeast must be replaced (start-up then takes about two days). The yeast is the same as employed in primary fermentation and forms a monolayer on the support matrix. Immobilisation takes place *via* the ion-exchange principle, which facilitates removal of yeast after cleaning. Spezyme GDC is rigid and non-compressible which enables it to be used under operating conditions of high flow-rate and pressure. The latter is important because it enables CO_2 to be maintained in solution. The carrier is inert and unaffected by any normal brewery cleaning chemicals (including caustic soda); thus the reactor can be incorporated into a standard CIP cleaning cycle. The two original reactors at Sinebrychoff are 1 m^3 in capacity and capable of conditioning 40 000 hl year^{-1}.

Using this maturation system, the brewery can respond to sudden increases in demand within one week. In breweries with traditional tank maturation facilities, sales increases have to be predicted!

Once the beer has been through the conditioning phase, it is filtered before being passed to the bright beer tank. Because of the volume of yeast produced after fermentation, beer is normally centrifuged before being introduced to the filters. Care has to be taken during centrifugation in order to prevent shearing and temperature increase in the product. The process is also noisy and costly.

Filtration serves three purposes:

(1) to sterilise the beer, *i.e.* to remove all yeasts and bacteria; a process called polishing;

(2) to remove all visible suspended non-biological material; called clarifying or roughing;

(3) to stabilise the beer further by preventing non-biological haze.

Before filtration, it is important to chill the beer to as low a temperature as is feasible so that as much chill haze and cold-break material as possible is deposited.

There are two different approaches to beer filtration, one of which is to remove all particles completely, such that the final product is sterile. This is carried out by a sieving mechanism and is called absolute filtration. Membrane filters are employed, normally with a pore size of no more than 0.45 μm. Because each particle removed effectively blinds the pore that removed it, the filters have a propensity to block quickly if turbid liquids are introduced to them. Thus, they can only really be used to provide a final polish to the beer after it has had most suspended material removed by prior, coarser filtration. Membrane filters are, however, used extensively to sterilise liquor and gas supplies in breweries.

For gross removal of yeast cells and other particles from beer a depth filter is required. The earliest examples were known as pulp filters and consisted of cellulose or asbestos fibres compacted into pads which were held in a frame; beer was then forced through the apparatus. After use, the pads could be dismantled, washed and reconstituted. By providing a latticework of interwoven channels, suspended particles are subjected to a very intricate pathway through the filter medium (the 'tortuous route') and they finally become trapped in blind alleys. Entrapped particles do not have to be larger than the mesh-size of the filter channels, so this is not merely a sieving mechanism. Adsorption, whereby particles are held by electrostatic forces, also plays a part in depth filtration and, again, particles smaller in diameter than the pore sizes in the filter bed can be removed. Most filter bed media carry a positive charge, to varying degrees. Eventually, depth filters become blocked by entrapped debris, but they have a far greater capacity than membrane filters. Pulp filters are still used occasionally these days, mainly to produce a preliminary rough filter. Their successors are plate and frame filters, whereby a series of framed textile sheets supporting the filter medium is held in an enclosed case. The filter sheets are loaded alternately with blank interspersing frames, with the beer being passed along the apparatus. These filters are mechanically simple, reliable and robust, and give a good quality filtrate; but, automation is virtually impossible.

Asbestos is rarely used these days and a number of other substances are commonly used to construct the filter, the principal ones being kielselguhr (a diatomaceous earth), Perlite (a siliceous volcanic glass) and silica hydrogel. Such compounds are referred to as filter aids, and to construct a filter (say, a kieselguhr plate filter) the following protocol is followed: the filter aid, in coarse form, is layered onto the cloth support by pumping a slurry over it under pressure. This bridges the larger pores in the support and is called the first precoat, or mechanical precoat. This is then overlaid by the second, or working precoat, which consists of finer-grade particles. Beer, mixed with predetermined amounts of the filter aid, can then be pumped to the precoated filter, the particle size of kieselguhr being the same as that used for the second precoat. This, in effect, extends the depth of the filter bed slowly; a process known as bodyfeeding. The rate of addition of the bodyfeed is critical, such that there is only a gradual thickening of the filter bed (which should be kept porous at all times during the filtration period). As more particles become lodged in the filter, more pressure is required to keep beer flowing. The filtration run comes to an end when all of the channels are blinded and all of the adsorption sites have been filled.

Modern depth filter technology involves sheets, leaves (vertical and horizontal), candles and cartridges, most of these types of filter being capable of being automated.

It is now common practice to incorporate a stabiliser, such as PVPP (polyvinylpolypyrrolidone), into the filter aid, thus aiding polyphenol removal.

Filtered beer passes to the bright beer tank from where it can be kegged, bottled or canned, the last two types of container being referred to as small packs. For large amounts, bright beer can be pumped to aluminium or stainless steel cellar tanks (up to 1000 litre capacity), which are sometimes equipped with a plastic lining into which the beer is situated. Tanks can be moveable, in which case they are filled in the brewery, or they can be fixed in the cellar of the outlet concerned, in which case the beer will be tankered.

Kegging

Kegging is the placing of bright beer into a stainless steel or aluminium container that only has a single aperture at one of the chimb ends. The aperture has a neck, called the Barnes neck, into which is fixed a downtube, or spear, which protrudes into the keg itself. Beer is introduced into the keg through the spear and, once in trade, beer will be dispensed from it. The spear has four main parts:

(1) the head (or body) which is screwed into the neck of the keg;
(2) the stem;
(3) a valve sealing ring;
(4) a spring to keep the sealing ring tightly adpressed to the top of the keg.

At the top of the keg there is a tamper-proof security cap to prevent any adulteration of the contents.

Some of the original kegs were 'cask-shaped' and of traditional size (9 or 18 gallon), but these have evolved into straight-sided containers, with lipped chimb ends, which are conducive to stacking and automated handling. Modern kegs tend to be in metricated sizes (50 or 100 litres, or 11 or 22 gallon).

Keg beer has found favour with brewers and publicans because it is a consistent, stable product with a shelf-life far in excess of that exhibited by cask-conditioned beer (12 weeks, as opposed to 4 weeks). It also requires little expertise on the part of the landlord in terms of its handling and general treatment. The spread and popularity of this type of product probably coincides with the increased mobility of the general public, who want to drink their 'local products' even when located in other parts of the country (or, indeed, world!). The pre-eminence of certain beer brands is largely attributable to the facility to keg.

Modern kegging lines are highly automated, and microprocessor-controlled machines can carry out all necessary stages, from de-ullaging to filling, in a single cycle. Kegging equipment, and the way it operates, varies, of course, from brewery to brewery, but a typical regime is given in the following paragraphs.

Empty kegs arrive at the brewery and are palletised (if they have not arrived on pallets). They arrive at the first station on the machine where they are de-palletised and inspected for damage. Inspection can be either manual or automated, and particular attention is paid to faulty downtubes and the presence of foreign kegs (*i.e.* not compatible with the line). The keg is then washed externally as it proceeds through a tunnel of detergent sprays and high-pressure water jets. The security cap on the top of the downtube is removed and the keg then passes to a station where the downtube is tightened to a specified torque before being passed to the core of the plant. Here it will be internally washed, sterilised and filled at separate stations. Kegs are moved along the lines by conveyor or moving beams. The cycle on the core plant is likely to be:

(1) Station 1, de-ullaging. The keg is attached to a head which removes any remaining beer.
(2) Station 2, detergent wash. A series of rinses and a detergent wash remove any vestiges of beer and scour out beer-stone. The detergent is chosen to be compatible with the metallic nature of the keg. Regimes vary, but a typical cycle at this station would be:

—first rinse: liquor at 75 °C for 15 seconds;
—drain for 2 seconds;
—detergent wash at 70 °C for 25 seconds;
—drain for 3 seconds,
—final rinse: liquor at 78 °C for 15 seconds.

Since the next stage in the cycle is keg sterilisation, it is common to purge the final rinse water with steam, whence it is collected and can be used for first rinse liquor in a subsequent wash. This also raises the temperature of the keg in readiness for sterilisation.

(3) Station 3, sterilisation. At this head the cask is steamed at 120 °C for about 1 minute.
(4) Station 4, counter-pressuring. Carbon dioxide, nitrogen, or a mixture of both are injected into the keg in order to purge out any remaining steam or condensate. This also produces a counter-pressure in the container which reduces fobbing when beer is introduced at the next head.

(5) Station 5, filling. At the filling head, pasteurised (or sterilised) beer is inserted; slowly at first to discourage fobbing. As the counter-pressure falls, filling can be speeded up. Beer enters *via* ports in the spear, and gas is voided through the centre of the spear. Kegs are then passed through a washer to remove any beer around the neck, check-weighed to ensure adequate filling and finally product-coded (labelled).

The most modern kegging machines consist of two rotating carousels, one of which carries out the cleaning cycle before the kegs are passed to the second (filling) carousel *via* a conveyor, on which they are steam sterilised. Such a plant is capable of filling up to 1000 kegs per hour.

If beer in the bright beer tank has not been sterile-filtered, then it will be pasteurised before kegging to increase the shelf-life of the product. The equipment normally used is a flash-pasteuriser, which is basically a heat exchanger; with cold beer flowing in one direction and hot water in the other. There are three stages:

—the product is raised to 73 °C;
—then held in tubes to give a dwell time of 30 seconds;
—and the beer is then chilled before being kegged.

To maintain the dissolved CO_2 level in the beer a pressure has to be maintained in the pasteuriser.

The degree of heat treatment is measured in pasteurisation units (PU), one PU being the lethal effect on micro-organisms obtained by holding the product for 1 minute at 60 °C.

The temperature effect on microbes is exponential with increasing temperature, and the number of PUs per minute can be derived from equation (6.2),

$$PU \ min^{-1} = 1.393^{(T-60)} \tag{6.2}$$

where T is the temperature in °C.

Small Pack

Small pack beer volumes have increased considerably since the late 1970s, largely due to the trend in the population of drinking more at home rather than on licensed premises. It probably now constitutes some 35% of all beer consumed in the UK, which for the brewer represents high volume, but small margin, especially since the larger supermarket chains entered the market.

Originally, glass bottles were returnable to the brewery for re-use (called multi-trip), but latterly, non-returnable glass bottles (NRB, or single-trip) and PET (polyethylene terephthalate) bottles have become very widely used. The latter can also be used only once.

Glass bottles vary in size from 180–1000 ml, whilst PET bottles are normally of 2 litre size.

Modern bottling lines are highly automated and capable of a very high throughput at relatively low noise levels, the apparatus varying somewhat for each category of bottle.

Returnable bottles are normally transported in plastic crates, and on return to the brewery the first stage in the bottling cycle is de-crating after which the bottles are assembled and emptied of any contents. This not only means beer residue, but anything that is capable of being put into a beer bottle! Once clear of debris they are checked for damage, especially neck damage, before washing proper commences. There are two main types of automatic washer: (a) soak washers, where the bottle is immersed in detergent and rinse tanks; and (b) hydro washers, where sprays are used. Early machines were all of the soak type; more modern ones tend to have soak and hydro features. Washers can also be single or double ended; in the former, bottles enter and leave the machine at a single point.

A cleaning cycle for multi-trip bottles is something like:

—Stage 1, pre-rinse: to remove gross soiling and to pre-heat bottles. Water used at 32 °C.
—Stage 2, first detergent soak tank: to loosen labels and soften any remaining soiling; 1.0% caustic soda used at 50 °C.
—Stage 3, second detergent soak tank: to remove labelling and dissolve remaining soil; 2.5% caustic soda used at 60 °C.
—Stage 4, detergent wash: to remove any internal soiling; 2.5% caustic soda used at 72 °C.
—Stage 5, third and fourth detergent tanks: complete removal of labels and soil; 1% caustic soda used at 72 °C.
—Stage 6, detergent wash: internal and external jetting of bottles; 0.5% caustic soda used at 60 °C.
—Stage 7, external hot (40 °C) water rinse: to remove detergent.
—Stage 8, internal hot (40 °C) water rinse: to remove detergent.
—Stage 9, external and internal cold water rinse: to remove all traces of detergent and reduce the bottle temperature to near ambient.

Holding times vary, but for caustic soak tanks 5–7 minutes is normal. Glass bottles are, of course, subject to thermal stress, and they withstand it better when being warmed than when being cooled. As may be assumed from the above cleaning cycle, label removal is the biggest single problem. Cleaned, cooled bottles are then electronically inspected before filling.

Non-returnable bottles that arrive from the manufacturer are nearly sterile (and devoid of labels) when unpacked and should only need a jet of sterile compressed air, followed by an internal spray of sterile water before filling. They are composed of thinner glass and require more sympathetic handling.

During bottle filling it is imperative to exclude oxygen from the system and most fillers have three chambers aimed at achieving this; one to hold beer, one to hold CO_2 and a spent gas chamber.

To commence filling, a bottle is lifted up onto a filling head where it is purged with CO_2 which then goes into the spent gas chamber; a counter-pressure is formed which equilibrates with that in the beer chamber. The beer valve opens and beer slowly enters the bottle, displacing gas as it does so. When the base of the filling tube has been covered, the beer filling speed increases. When the appropriate volume has been introduced, a valve prevents any more gas escaping to the spent gas chamber, which means that no more beer can enter the bottle. The beer in bottle is allowed to stabilise and its pressure is lowered slowly to atmospheric before being outfed to a high-pressure water jet which causes the beer to foam. Foaming causes any residual oxygen to rise to the very top of the bottle where it is voided. After jetting, the bottle reaches the capper, or crowner, the most usual means of closure.

There are two types of filling tube: long and short. The former reaches to the bottom of the bottle and emits beer, whilst the short tube sprays beer down the inside of the bottle wall.

At the closure point, the bottle is raised to meet the metallic crown which is held magnetically in a die. Once contact has been made, an outer sealing ring brings down the skirt on the crown around the neck of the bottle and a crimping occurs to form a seal. The lining of the crown, which consists of polyethylene, produces an airtight seal between metal and glass. Some crown linings are oxygen-scavenging. Sealed bottles then pass through a tunnel pasteuriser to ensure the microbiological stability of the product. Bottles are transmitted through several zones of jetted water, of varying temperature, such that they are gradually heated to 60 °C, held for a set period and then gradually cooled. A typical regime in a tunnel pasteuriser is as follows:

(1) a pre-heat zone, where the product is raised from 5 to 35 °C;
(2) a super heat zone, where the product is heated to 60 °C;
(3) a pasteurisation zone: the product is held at 60 °C, where the time spent here is dependent upon number of PU desired;
(4) first cool zone: the product is cooled to about 50 °C;
(5) second cool zone: the product is cooled to 30 °C;
(6) final cool: the product is cooled to 20 °C.

The degree of pasteurisation required is, to a large extent, dependent upon the nature of the product and its pre-bottling treatment; the better the microbiological quality of the beer, the less back-up heat treatment needed. In the UK, recommended levels of dosage are higher (20–30 PU) than in the US (where it can be as low as 5 PU).

After pasteurisation, bottles are electronically checked for content, labelled (called dressing) and re-crated, if multi-trip. Single trip bottles are either boxed or shrink-wrapped in cardboard trays.

Beer destined for canning will have followed the same overall course as for bottling, *i.e.* it will be in a bright beer tank. Cans are, of course, one-trip containers, and canning lines, together with non-returnable bottling lines, are much easier to control than multi-trip bottling lines. This is because the outfeed end is not linked to the infeed (because no crates are involved).

Originally, cans were three-pieced (body and top and bottom ends), but nowadays they are formed from steel or aluminium and arrive at the brewery in two parts; a cup-shaped body and a top end (which contains the aperture for dispense, *e.g.* the ring-pull mechanism). The body shape is produced by a process called drawn and wall ironing (DWI), whereby the strip metal is lubricated with oil and then passed through a cupping press which produces shallow cups. Each cup is then rammed through a series of tungsten rings which causes the cup walls to elongate and become thinner. The tops are then trimmed to size and the bodies washed (to remove oil) and dried before being lacquered. The can wall is now printed, where the design is introduced lithographically or in a process whereby a reverse-printed label is wrapped around the can which is then varnished and heated in an oven. The ink from the label vapourises onto the lacquer of the can; a process called reprotherm. All printing has to be inserted once the can body has been rounded.

Steel cans are constructed from a special grade of low-carbon steel which is coated on each side by a thin layer of tin; this arrangement protects the steel from corrosion. Aluminium cans, which comprise about 60% of the market in total, consist of the element alloyed with magnesium and manganese to give greater strength and ductility; different thicknesses and strengths are used to make the bodies and the ends.

Prior to filling, cans are placed upside down and rinsed with sterilised water. Modern fillers are of the unflooded beer chamber type, where the beer container is half-filled with beer and covered with inert gas (CO_2 or N_2). Cans are introduced to the filling point and have CO_2 or N_2 pumped into them. When the can is counterpressured, beer flows into it at low velocity (to prevent fobbing) *via* a short tube positioned down the side of the can. A ball valve terminates beer flow when the correct filling level has been reached. Cans are then transferred to the seamer where they are surface blown with inert gas to disperse any bubbles; called bubble-breaking. A filled can is then either raised to meet the top end, or, as in more modern machines, the top is lowered down to meet the can body. As the two components meet, CO_2 is blown over the beer surface for final removal of any air; called under-cover gassing. The can top is then joined onto the filled body by means of a two-stage operation involving rotating chucks which produce a flat double seam. The can body will have been flanged during production so that the top can be received.

The closed cans are inverted and checked for correct filling level by means of a gamma radiation level detector. They then pass to a tunnel pasteuriser where a normal treatment is:

—preheat at 46 °C for 1 minute;
—superheat at 62 °C for 3–4 minutes;
—pasteurise at 60 °C for 15 minutes;
—precool at 46 °C for 2 minutes;
—cool at 32 °C for 2 minutes.

This provides approximately 20 PU for the product. Cans are checked post-pasteurisation for correct content levels and are then coded on their base, by ink jetting, with the 'best before' date.

Packaging follows and this normally consists of either shrink-wrapping in trays, or covering groups of cans (4 or 6) with a stretched plastic shroud, called a high cone.

Canned beers have a much higher CO_2 content than draught beers (4.5 g l^{-1}, as opposed to 2 g l^{-1}), principally to ensure oxygen exclusion and to produce an internal pressure against the decreasing rigidity of ever-thinner can walls. In recent years, much work has been carried out with the aim of making canned products more akin to draught ones, especially in terms of taste and appearance. This means producing a stable, creamy head and a lower dissolved CO_2 level. The result has been the development of the 'widget', a device inserted into the can itself which causes a high velocity movement of gas (usually N_2) through the beer when the can is opened. Beer in widget cans is nitrogenated to around 20 ppm, and it is recommended that the product be stored in a refrigerator for a few hours before use, to increase the solubility of the gas. On opening, the widget induces shearing and the evolution of gas in minute bubbles, which float on the beer and produce the desired tight, creamy head. Nitrogen can either be introduced by gaseous supersaturation (dissolving it into chilled beer under pressure) just before filling, or by injecting liquid nitrogen into the filled can immediately prior to seaming. Cans containing widgets must be oversized relative to the volume of beer that they contain; this is to accommodate not only the widget itself, but gas and a sufficient headspace. A normal sized can is 500 ml (to hold 440 ml beer). Widgets are normally plastic objects of varying design, but Bass have developed an aluminium version which makes can re-cycling far easier.

PET bottles, commonly called plastic bottles, are normally of 2 litre size in the drinks industry, although other sizes are available. They are used mainly for soft drinks; only a few breweries now favour them. Because of their inherent slight porosity they are lined with ethylvinyl alcohol (EVOH) or polyvinyl dichloride (PVDC) to prevent CO_2 escape and oxygen ingress. Bottles enter the brewery pre-formed and are purged with inert gas and then rinsed with sterile water. Bottling lines are similar to those for glass bottles, but beer must be flash-pasteurised before filling because PET will not withstand pasteurisation temperatures; thus, filler, bottles and caps must be sterile. After filling, bottles are jetted before being capped, both filler and capper possessing sanitising jets. When initially sealed, bottles are non-turgid and have to be firmed up before they can be labelled. This is done either by passing them through a tunnel at 40 °C, or by inverting and twisting the bottles in order to promote CO_2 breakout in the bottle. PET bottles are usually presented in shrink-wrapped trays.

NITROGENATED BEERS

During the 1940s, Guinness, experimenting with the use of nitrogen in the packaging of their stout products, found that small amounts of the gas gave an increase in the texture (creaminess) and durability in the head (foam) of the beer after dispense. Guinness had, in fact, patented a two-part 'keg' for nitrogen dispense in 1932. In those days, beers were not 'brewery conditioned' and were, in effect, being dispensed by top gas pressure, and it was not until the evolution of kegging in the late 1950s/early 1960s that nitrogen was used to any extent as a means of dispensing keg stouts. Just as breweries had sought to utilise surplus CO_2 (from fermentation) for commercial purposes, *i.e.* by carbonating beer, so the increased use of nitrogen in certain large breweries (*e.g.* as a counter-pressure gas during packaging and for deoxygenating water during high gravity brewing) led to them looking for other ways of using the gas.

Nitrogen is attractive for a number of reasons, probably the most important being the fact that by means of its very structure, *i.e.* diatomic and triple-bonded, it is unreactive chemically (inert), certainly in terms of anything that it is likely to come into contact with in the brewery. It is also relatively insoluble in water, especially when compared with CO_2, which is around 100 times more soluble. In brewing terms, it is non-toxic, odourless and tasteless. Because of its lower density, it is also less expensive than CO_2 (it is reckoned that it is approximately 60% cheaper to use). This is based on the fact that, under STP conditions (standard temperature and pressure), one tonne of N_2 will liberate 8400 hl of gas, whilst the same weight of CO_2 liberates only 5300 hl.

It is generally accepted that CO_2 gives an acidic, undesirable, prickly taste to beer, especially when any pressure is involved to impart it. In order to imitate traditional, cask-conditioned ales, which would only have small amounts of CO_2 dissolved in them as a result of secondary fermentation, some breweries have managed to persuade a certain section of the consumer population that nitrogenated (nitrokeg) products are 'real ales' dispensable from kegs or cans. Nitrogen is invariably mixed with CO_2 in a predetermined ratio, dependent upon the beer type (usually 60 : 40 or 70 : 30 CO_2 to N_2). It is interesting to note that there are no generally accepted standard units of definition for N_2 in the brewery. Industry reports indicated that in 1998 some 25% of all ales and stouts were dispensed by mixed gas, and that this figure was likely to rise to 50% within the next few years.

Fitch[3] provides a good account of the developments in the use of nitrogen within the brewing industry.

HIGH-GRAVITY BREWING

The technique of high-gravity brewing found considerable favour with certain larger breweries, particularly in the US and Canada where more beer is produced by this method than by conventional means.[4] It has even been stated recently, that 'The use of high-gravity brewing techniques is essential for the future economic viability of the brewing industry'.[5] Whatever the merits, this is a process whereby worts of a higher than required gravity [normally in the range 1048° (12 °Plato)–1072° (18 °Plato) – for a definition of °P, see pages 182–183] are produced and then subsequently diluted with water until the desired strength is obtained. In theory, breaking down with water can occur at any of the following stages of the brewing cycle:

—on casting out the copper;
—pre-wort cooling;
—post-wort cooling;
—during fermentation;
—after fermentation (most common);
—during maturation;
—pre- and post-filtration.

The exact point of dilution will determine the quality of water to be used. All water should be of brewing quality and pH adjusted. For kettle additions, the water should be microbiologically sound, whilst for all post-kettle dilutions it must be sterile-filtered. Water for post-fermentation dilutions should, in addition, have a dissolved oxygen (dO_2) level of less than 100 ppb.

The majority of breweries that resorted to this technique on a commercial basis did so in order to avoid excessive expenditure on additional brewing capacity, particularly in terms of wort production and wort boiling. These facets apart, a number of advantages have been claimed for this concept of brewing. They may be summarised as follows:

(1) reduction of the amount of water used in the brewhouse;
(2) reduction in energy costs (heating and refrigeration);
(3) reduction in labour costs (*e.g.* cleaning down);
(4) reduction in effluent costs;
(5) can use higher proportions of unmalted carbohydrate adjuncts;
(6) can get more alcohol per unit of fermentable extract;
(7) beers have improved physical and flavour stability;
(8) finished beers have a 'smoother taste';
(9) can get much more flexibility in product portfolio; several beers, of widely differing gravity, colour and hop character can be produced from one batch of wort. This is especially true now that pre-isomerised hop extracts and natural colouring materials, derived from brewing raw materials, are freely available.

The known disadvantages of this form of brewing are documented as follows:

(1) water used for dilution must be sterile and deoxygenated;
(2) the mash will be more concentrated (increased concentration of carbohydrate to water) and this will give decreased efficiency in terms of extract;
(3) hop utilisation rates in the copper are lower;
(4) foam stability is decreased, leading to beers with poor head retention;
(5) yeast performance can be impaired, owing to higher osmotic pressure of the worts (initially) and high levels of alcohol (during the later stages of fermentation);
(6) yeast flocculation characteristics are affected; many top fermenting strains revert to being sedimentary.

Experience has also shown that it is far more difficult to undertake all-malt high-gravity brewing, and this is principally because of overloading of mash/lauter tuns and subsequent wort-separation problems. The technique lends itself best to high adjunct recipe beers, especially where liquid adjuncts can be introduced into the copper. This relieves mash tun loading. Beers with low hop rates are also easier to create by this method.

In the last few years a plethora of research work has been undertaken with a view to understanding the reasons for the above-mentioned disadvantages. Particular attention has been paid to head (foam) formation and retention, probably one of the main negative features of high-gravity brewed beers. There are many foam promoting and stabilising compounds in beer, the most important being proteinaceous. It was originally thought that certain sized proteins (8, 10 and 40 kD) were most effective in stabilising foam, although it is now thought that that the degree of hydrophobicity of the proteins is more significant than their size.[5] The researchers on this work showed that hydrophobic proteins are lost during the course of the brewing process, and more rapidly so in higher gravity wort brews. It was also evident that the actual production of high-gravity wort did not extract as much hydrophobic protein as could be obtained from conventional gravity mashes. The use of modern mash filters for wort separation seems to increase hydrophobic protein extraction, especially if the grists have been prepared by use of a hammer mill, which is known to give grists that yield better protein degradations during mashing. Mash filters also promote foam stability by being able to remove haze particles and fatty acids from wort. The latter, especially, are foam-negative entities.

Extensive work, aimed at improving yeast performance during high-gravity brewing, has been carried out in the Research Laboratries of the Labatt Brewing Co., London, Ontario. In 1987, D'Amore *et al.*[6] demonstrated the effect of osmotic pressure on the intracellular accumulation of ethanol during fermentation. They found that there was an accumulation of intracellular ethanol during the early stages of wort fermentation by *Saccharomyces cerevisiae*, and that as fermentation proceeds, intracellular and extracellular levels became similar. An increase in wort osmotic pressure, from 10 to 20 °P, caused an increase in intracellular ethanol concentration and glycerol production without any detrimental effect on cell growth and fermentation, but did result in the production of more extracellular ethanol per unit time.

In subsequent work,[7] the same group investigated such parameters as yeast pitching rate, fermentation temperature and wort oxygenation in relation to the fermentation of a series of high-gravity worts (up to 34 °Plato), prepared by supplementing a standard 16 °P wort with high-maltose corn syrup. The small-scale (30 litre) experiments showed that by selecting the correct combination of initial oxygenation, yeast pitching level and fermentation temperature, a beer of 25 °P could be produced which compared favourably with one emanating from production wort gravity. Under the prescribed conditions the 25 °P wort underwent complete attenuation with no increase in fermentation time. Worts above this gravity showed a decrease in the rate and extent of fermentation, and so, at present, very high-gravity brewing may not be feasible.

Most of the brewing information outlined in this book pertains to the production of 'standard' ales and lagers. Many other beer styles exist, of course, and it is pertinent to mention three of them.

LOW ALCOHOL BEERS (NABLABS)

The definition of non- and low-alcohol beers differs throughout the world. In the UK, non-alcohol beer refers to a product containing less than 0.1% alcohol by volume, whilst low-alcohol beers have a higher alcohol content, normally between 0.5 and 1.5% v/v. Light (or lite) beers also have reduced alcohol levels, normally between 1.6 and 2.5% v/v. In the US, beverages containing less than 0.5% alcohol by volume cannot be described as beer.

Increased awareness of alcohol-related medical conditions, coupled with stricter drink–driving legislation, have ensured a market for such products and there is now a great social appeal for a low- or non-alcohol product that actually 'tastes and looks like beer'. Brewers have known for many years that to produce such a beverage is difficult, especially when it is realised that the flavour contribution made by alcohol to a beer has to be compensated for by something else. Much research has been carried out into obtaining the correct flavour profile, colour, aroma and head retention capacity for these styles of beer.

There are two main types of process used to obtain reduced alcohol beers:

(1) limited, or checked, fermentation;
(2) post-fermentation techniques, *i.e.* the removal of alcohol from normally-fermented beer of conventional strength.

The earliest low-alcohol beers were all produced by the first category of process, which, by definition, normally led to sweetish products as a result of unfermented wort sugars. Because of this inherent problem, there was a gradual progression to post-fermentation processes, and it is now evident that the operation and control of such processes is much easier, thus enabling the production of a more consistent beer. Certainly, the products produced by alcohol removal processes are now generally regarded as being more akin to conventional beer.

(1) *Limited, or checked, fermentation processes.* Techniques used in this category are:
  (a) Checked fermentation (limited fermentation). This is effected by arresting the yeast growth of a normal fermentation by rapid cooling to 0 °C. This is only normally applied to lager-style beers, where fermentation temperatures do not usually rise much above 8–10 °C. Worts normally originate at around (say) 1040 °OG (10 °Plato) and these are fermented until the alcohol level has reached 2.5% v/v. The beer is then immediately chilled to 0 °C, whence the yeast is removed. The beer has to be heavily pasteurised because of its inherent microbiological instability.
  (b) High temperature mashing. By mashing malt (or malt and corn syrups) at elevated temperatures (75–80 °C), modified worts are produced, which, if fermented to about 45% of their normal level, yield beers of around 2.5% ABV (for a

definition of ABV, see pages 183–184). Again, such beers show instability and have short shelf-lives.

(c) Use of a yeast with limited fermentation capacity, such as *Saccharomyces ludgwiggi*. Unlike *Sacch. cerevisiae*, this species cannot ferment maltose from conventional worts. Sucrose, glucose and fructose are metabolised, however. Since some 40% of the fermentable content of a normal brewer's wort consists of maltose, one can appreciate the theoretical possibilities. For example, by adjusting the mash components it is possible to produce worts with a combined sucrose–glucose–fructose content of only 15%. After fermenting such worts, beers with an alcohol content of only 0.5% can be produced. Unfortunately, the unfermented maltose enhances the sweetness of the finished product to sometimes unacceptable levels.

(d) The 'Barrell' process. A patented method that involves fermenting two sets of worts: one of normal OG (say 1040°); one of low OG (say 1012°) and blending the resultant beers. The worts should have the same pH, colour and bitterness levels. During fermentation of the stronger wort, all of the CO_2 and volatiles are fed into the vessel containing the fermenting weak wort. The resultant beers are blended in predetermined proportions to yield final products of varying alcohol content. Thus, by blending in the ratio 1:2 (1040°:1012°), one produces a beer of 2.4% ABV, whilst blending 1:4 (1040°:1012°) results in a 0.9% ABV product. The patent claims that, because volatiles are not lost, beers with conventional flavours are manufactured.

(e) The 'cold-contact' process. Developed by Schur,[8] this process essentially involves 'fermentation' at low temperature. In fact, conventional worts (normally acidified to pH 4.0) are held at around -0.5 to $-1.0\,°C$ and placed in contact with a yeast culture from an active fermentation. The yeast should be washed free from traces of alcohol and the contact time should be for a period of around 48 hours. Yeast pitching rate is critical (30×10^6 cells ml^{-1} wort) and the whole fermentation must be continually roused; this is usually by recirculating pump. After the alloted contact time, the yeast is removed and the beer is immediately stabilised with PVPP. Beers with an ABV as low as 0.05% can be produced by this method.

(f) Spent-grain process. Spent grains from the mash of a high-gravity brew are digested by soaking in hot water and an extract is made. This is then boiled with hop extract for 1–2 hours to give the required α-acid level. The 'hopped wort' is then fermented; the beers produced are normally in the range of 0.5–1.5% ABV.

(2) *Post-fermentation techniques.* There are four main methods by which normally brewed beer is de-alcoholised. After alcohol has been removed the product is normally blended with a mixture (or cocktail) of aroma and flavour components in order to achieve 'normal' beer characters.

(a) Distillation. This can be performed at ambient temperature or under vacuum. In the former process, clarified beer is slowly introduced into a copper which contains boiling water. A boil is maintained until the beer volume is reduced by about 30%. Beers can be obtained down to 0.5% v/v alcohol by this means, but, unfortunately, the prolonged heating affects hop constituents and the beers are mal-tasting unless rectified.

Vacuum distillation is not so deleterious to beer flavour since lower temperatures are employed. It is also possible to remove the desirable flavour components before the alcohol is removed – a process called de-esterification. These can be reintroduced to the beer once the alcohol has been removed. In theory, it is possible to remove all traces of alcohol by vacuum distillation.

(b) Evaporation. A very flexible technique which can yield beers with varying low-alcohol levels. Both rising and falling film evaporation methods have been used, operating usually at 30 °C and 30 mm Hg. Flavour components are not usually removed prior to the evaporative stage and so heat treatment must not be excessive. To avoid thermal stressing of the beer, the temperature difference between the incoming beer stream and the steam evaporating alcohol from it is as small as possible. The process can be operated on a batch basis or can be continuous if desired. Water and CO_2 are inevitably removed as well as alcohol, and what actually constitutes the product stream from the evaporator is a de-alcoholised beer concentrate. This has to be thinned with carbonated water (to replenish water and CO_2) and then blended with a mixture of flavour compounds to re-establish the overall beer characters. ABVs can be reduced to as low as 0.01%.

Both evaporation and distillation use phase separation of the alcohol from the initial beer.

(c) Reverse osmosis. This is the high-presure filtration through a semi-permeable membrane of a normally-brewed beer. The relatively small size of the ethanol molecule enables the technique to be used. There is very little detrimental effect on flavour components because the system operates at ambient temperature, or below. In practice, the beer is diluted with deoxygenated water before reverse osmosis, to compensate for the loss of water to the permeate and to prevent clogging of the membranes. Some small molecular weight flavour compounds may also be lost to the permeate and so some adjustment may have to be made to the de-alcoholised beer.

The degree of reduction in alcohol content can be regulated by the choice of membrane; in practice, it is difficult and expensive to reduce ABVs to below 0.5%.

(d) Dialysis. A technique developed in the 1980s which is now regarded as being the most satisfactory method of removing alcohol from beer; the end-products are certainly more generally acceptable. The process operates at cellar temperature and near ambient pressure, the pressure difference across the dialysis membrane being around 0.1 bar. Such operating conditions do not have an adverse effect on final beer flavour.

The dialysis membrane is normally constructed from pure cellulose and is held in a module. Beer and the dialysate (which usually consists of potable water with a small percentage of the beer being de-alcoholised) flow counter-current across it. The concentration gradient causes alcohol to diffuse across the membrane. Diffusion is, in fact, a two-way process, the direction of passage of molecules being determined by concentration gradients across the membrane. The extent of alcohol reduction is controlled by the flow rates of beer and dialysate and their residence time in the dialysis module. Ideally, the dialysate should contain dissolved CO_2 in order to prevent excessive loss of the gas from the beer stream. Some desirable small molecular weight substances can be lost to the dialysate, but these can be recovered later on after the alcohol has been removed by vacuum distillation. After alcohol removal and recovery of essential substances, the dialysate is, in theory, re-usable.

A totally different concept for producing low-alcohol beer was introduced in 1988 by the Bavaria Brewery, Lieshout, the Netherlands, who employed immobilised yeast to produce alcohol-free and low-alcohol beers in very short periods of time; one of the products was known as *Bavaria Malt Drink*. Brewing takes place in bioreactors containing Cultor Spezyme GDC support medium. The worts used are 'modified' and these are clarified and passed through the reactor. Fermentation in the reactor is controlled by the modifying yeast metabolism. The beer produced is diluted, carbonated and stabilised before being bottled and pasteurised. According to how the plant is operated, beers from 0.05 to just over 1.0% ABV can be produced, and beer can be in bottle from 2–5 days after production commences.

ICE BEER

Bottled ice beers became something of a vogue in the early 1990s, probably mainly due to extensive marketing exercises! It has been known for a long time that if a beer is frozen, ice crystals form, which, in effect, constitutes the removal of water; other beer constituents remain in the liquid phase in more concentrated amounts. This is called freeze concentration and is the manner in which the strong, speciality German eisbock beers, for example, are produced.

According to Bamforth,[9] the actual evolution of the new-fangled ice beers emanated from technology originally designed to alleviate the problem of transporting finished bulk beer over large distances. Certain North American breweries looked at the idea of freeze concentrating their products, transporting in bulk and then reconstituting them on-site to sale strength before packaging. This proved ultimately to be non-viable.

As a result of perseverance, and the desire to utilise some of the technology evolved, the Labatt Brewing Company in Canada introduced 'ice brewing' in 1994[10] with some success. Fermented beer is centrifuged to remove yeast, cooled to -1 to $-2\,^{\circ}\mathrm{C}$, and then pumped into scraped surface heat exchangers (SSHE). The heat exchangers cool the beer to between -3.5 and $-4.5\,^{\circ}\mathrm{C}$, whence minute ice crystals form. An SSHE is used because it will not block with ice crystals. This mixture is further mixed with large ice crystals and the conglomeration is continuously agitated at the specified freezing temperature, the final ice concentration being anything up to 25% of the slurry volume. The process is continuous and there is a set contact time between ice and beer during the operation, after which ice-free beer is extruded into conditioning tanks. The low temperatures involved encourage early formation of haze, which is filtered off, resulting in a product with a consistently brilliant appearance. It is also claimed that such beers are more mellow and possess enhanced stability.

WHEAT BEERS

These unique, naturally cloudy beers have been part of the Belgian and German brewing cultures for many years and in certain regions were considered to be the preserve of the nobility. A wide variety of wheat beers exists, all of which have totally different characters from ales and lagers. Most have biscuity or spicy flavours, the latter being mainly attributable to vinyl guaiacol which emanates from ferulic acid (4-hydroxy-3-methyl cinnamic acid). In most instances it is the mashing regime that governs the final characters of a beer, *e.g.* to produce optimum amounts of ferulic acid (and hence the desired spiciness) one should employ a mash temperature of 47 °C at a pH of 5.3.

Because wheat malt forms a major proportion of the grist, and because of the intended nature of the final product, a number of facets of the brewing cycle are necessarily different. Mashes tend to be thinner to overcome the likelihood of setting and poor wort run-off, although modern mash filters can alleviate the problems. Boiling times are often shortened in order to minimise protein coagulation. As a consequence there are lower levels of DMS (dimethyl sulfide) loss during coppering, but this does not affect the final beer flavour profile as much as it would do for ale production. Fermentation vessels need to be high-sided to provide plenty of headspace during turbulent fermentations.

The beers themselves have inherent problems:

—they are microbiologically unstable, especially due to bacterial contamination promoted by low bitterness levels, high pH, high levels of protein and yeast autolysis;
—wheat beers are prone to gushing;
—they have a very substantial, foaming head which is likely to cause dispense problems.

The variety of wheat used during mashing will, of course, influence the final flavour of the beer, and until recently the malting of wheat was nowhere near as well studied as barley malting. Relatively few wheat varieties exist purely for malting for the brewing industry.

Following the relative success of some continental imports in the last few years, there has been some interest in wheat beer production in the UK. Initially, the wheat varieties used were primarily intended for the food industry, but the introduction of the Atlantis variety has provided British and continental brewers with a sound malting wheat. It has a low protein content and gives a good extract. Of particular benefit is its resistance to fungal spoilage, especially that due to *Fusarium*, a major cause of gushing. The variety has been used in Germany since the early 1990s and it is estimated that over 75% of Bavarian Weißbiers are mashed from Atlantis malt.

GRAVITY AND BEER STRENGTH

Until 1 June 1993, UK beer duty payable to HM Customs and Excise was calculated on the basis of the original gravity (OG) of worts prior to fermentation. In effect, this represented a tax on sugar rather than on alcohol. The OG measurement would be taken in an accurately-gauged fermentation vessel prior to the addition of yeast (or the commencement of fermentation), and a dip would be taken to ascertain the volume of wort in the vessel; this would be calculated from standard tables. Gravity measurements were made with standard saccharometers and the results expressed in degrees saccharin at 20 °C By this method, the gravity of pure water was taken as 1000° and the OG of a standard bitter would have been of the order of 1036° (which essentially means that there are 36 fermentable parts per 1000 parts of water present before fermentation commences). Although this was an approved method for establishing gravity, the very nature of wort (*i.e.* sugar and other substances dissolved in water) really means that at least two analytical methods should be used for total accuracy. OGs were measured to 0.1 ° saccharin (*e.g.* 1036.5°) although duty was payable on a gravity band, *i.e.* a beer of 1036.1° would incur the same amount of duty as one of 1036.9°.

The most accurate method of assessing wort strength is to measure specific gravity (SG) by means of weight measurement of a known volume in a specific gravity bottle. The SG of a sample is defined as the weight of the wort sample in relation to an identical volume of distilled water. A wort with a specific gravity of 1.040 approximates, in practical brewing terms, to an OG of 1040°. These days, most gravities are expressed in degrees Plato (°P). Dr F. Plato devised a scale which directly relates wort strength to the amount of sucrose in solution. By definition, 1 °P is equivalent to 1 g of sucrose dissolved in

1 ml of distilled water. Thus, again in practical terms, a wort sample with an SG of 1.040 (OG 1040°) represents 10 °P.

As fermentation proceeds and sugars are converted into alcohol, the gravity of the beer falls. Gravity measurements taken during this period are referred to as present gravity (PG) or actual gravity (AG). PG or AG can be used to monitor worts emanating from the mash tun, or, indeed, any other stage during beer manufacture. By definition, PG is defined as the specific gravity of a beer (or wort) sample multiplied by 1000, minus 1000 (and expresed in degrees saccharin at 20 °C). Final gravity, or finishing gravity, (FG), is the gravity at which the brewer will deem to be the end-point of fermentation, either in the fermenter or conditioning tank. The lower the FG, the more alcohol will have been produced from the worts of the stated OG. The brewer's decision concerning the required FG will determine the beer style.

Any unfermented wort contributes to the sweetness of the beer, and this can be adjusted by adding priming sugars to the cask or prior to packaging. The addition of priming sugars actually contributes more to sweetness than any unfermented wort residues.

All yeasts have an attenuation limit (AL) which is a measure of how much fermentable material they can remove from wort during the course of fermentation. The residual sugar is normally dextrinous and represents the difference between a final gravity of, say, 1008° and the theoretical point of total removal of fermentables, which could be 1000°. AL is, therefore, normally a positive figure, but, in certain circumstances, whereby there has been a total removal of fermentables, it may be a negative value. This is because the specific gravity of alcohol (0.791) is less than that of water.

Apart from direct measurement, there are a number of other ways of calculating the amount of alcohol produced during fermentation. One of the earliest methods was based on the proof spirit system, whereby it is calculated that for every 4.3° fermented, 1° proof spirit is produced. By definition, proof spirit contains 57.06% ethanol in water, a mixture which, when poured onto gunpowder and lighted, will allow ignition. Thus, 1° proof spirit equates to 0.5706% alcohol by volume (ABV). For a beer of OG 1040° fermented down to 1009° FG, the ABV can be calculated using equation (6.3) as:

$$\frac{40 - 9}{4.3} = 7.21° \text{ proof} = 4.11\% \text{ ABV} \qquad (6.3)$$

This does not, however, tell the whole story because, during fermentation, compounds are produced from, say, glucose, other than just ethanol, water and CO_2. So, for an approximate calculation of ABV it is more appropriate to use the expression,

$$OG - FG \times 0.129 = ABV \qquad (6.4)$$

which, for our 1040° beer fermented down to 1009°, as above, gives equation (6.5):

$$31 \times 0.129 = 3.99\% \; ABV \qquad (6.5)$$

On 1 June 1993, the UK came into line with other EEC member states and excise duty became payable on the alcoholic strength of a beer when it left the brewery gate (*i.e.* when sold). The new method of duty payment necessitates accurate measurement of ABV and the recommended IOB method employs distillation (Method 9.9).[11] Other precise methods are available, such as gas chromatography. Results are expressed in terms of %ABV to the nearest 0.1%.

BITTERNESS

The bitterness of beer is measured by an international method (IOB Method 9.16[11]) which involves spectrophotometry. An iso-octane (2,2,4-trimethylpentane) extract of de-gassed, acidified beer is placed in a 10 mm silica cell and its absorbance measured at 275 nm against a pure iso-octane standard. The absorbance so obtained is multiplied by 50 to give Bitterness Units (BU).

Although this is the standard method, it is only applicable to beers that have had bitterness imparted by boiling wort with hops, *i.e.* if isomerised hop extracts have been used post-fermentation, then there may be considerable discrepancy between measured bitterness and organoleptically-determined bitterness. Results are also only valid if the beer does not contain saccharin, salicylic acid, sorbic acid or *n*-heptyl-4-hydroxybenzoate, all of which are extracted by iso-octane and absorb at 275 nm.

COLOUR

The international method for assessment of beer colour (IOB Method 9.1[11]) involves spectrophotometric measurement at a wavelength of 430 nm; the sample is held in the customary 10 mm cell. The absorbance obtained is multiplied by 25 to yield European Brewery Convention (EBC) units of colour. The method can be applied to all beers (or worts) as long as they have been filtered prior to spectro-photometry.

Beer colour can vary from 0–350 °EBC, the normal range being 15–40 °EBC for standard British bitters. Pilsner lager normally has a colour of around 10 °EBC, whilst at the other end of the scale, stouts can be of the order of 90 °EBC.

The main contributions to beer colour are made by melanoidins emanating from malt and caramelisation products produced during boiling. Colour can be enhanced by artificial addition of caramel or another food colouring agent.

Caramels have been used as colour intensifiers and adjusters for over 100 years; they provide a means of adding reliable, predictable colour and flavour. They were originally produced as an offshoot of the brewing-sugar industry and the technology of caramelisation is now well known. Worldwide, caramel accounts for some 95% by weight of the total usage of food colourings. Being food additives, caramels must conform to the rigid standards laid down by statutory bodies; this applies particularly to its Acceptable Daily Intake (ADI).

Because of the wide range of uses, caramels have been developed to satisfy a variety of stabilities, and colour and flavour profiles. There are a number of ways of producing caramels, but they all involve heating a carbohydrate source (glucose syrups or sucrose) with food-grade reactants.

The WHO and EEC accept four classes of caramel:

—Class 1. That prepared by the controlled heat treatment of carbohydrates with, or without, the presence of food-quality acid or alkali. They are used in spirits and liqueurs.
—Class 2. Prepared by the controlled heat treatment of carbohydrates with caustic sulfite. Used in aperitifs (*e.g.* vermouth).
—Class 3. As for Class 2, but employing ammonia instead of caustic sulfite. Used in brewing, baking and the manufacture of meat products.
—Class 4. Prepared by the controlled heat treatment of carbohydrates with ammonium and sulfite-containing compounds. Used in soft drinks.

The ammonia caramels (Class 3) are particularly suitable for brewing because of their stabilty and colouring power. The colour intensity of most brewing caramels falls in the range 32 000–48 000 °EBC.

The high level of stability of brewing-caramels is derived from the fact that their isoelectric points are in the region of pH 6.0–6.5. At pH levels below this they carry a positive charge and so, in beers, where the pH would normally be around 4.0, they carry a positive charge which does not conflict with positive charges on other colloidal beer constituents. If this were not the case then there would be a danger of precipitation and haze formation upon caramel addition.

Brewing-caramels also have an inherent microbiological stability and do not require addition of preservatives before being stored in bulk.

One of the natural by-products of ammonia caramel manufacture is 2-acetyl-4-tetrahydroxybintyl imidazole (THI), and regulations insist that this compound should not be present at levels above $10 \, mg \, kg^{-1}$.

FOAM

One of the perceived features of 'a good pint' is the foam, or head, on top of it. To the individual drinker, the importance of the head varies in different parts of the world, or indeed, in the UK at least, there is a regional variation in the significance attributed to the head. In northern Britain a tight, creamy head is an all-important feature of a draught beer; this feature is not necessarily deemed to be a requirement in southern counties.

Beers dispensed from bottles, cans and kegs, where some degree of pressurisation has been employed in processing, will be supersaturated with gas (CO_2, N_2 or both). This means that they hold more dissolved gas than they should theoretically do for the existing temperature and pressure. Traditionally, before the widespread use of N_2, bottled and canned beers would normally have contained around 2.5 volumes of CO_2 dissolved in them. Even cask-conditioned beers exhibit a small degree of supersaturation and it is a fact that a beer must be supersaturated before a foam can be established, *i.e.* before the gas in solution can escape in the form of bubbles.

The initiation of a bubble requires a nucleation site, typically a rough surface. A liquid (beer) will not uniformly wet a surface with which it is in contact; it will bridge over any minute indentations on that surface (which will be present, no matter how smooth the surface appears to be) and trap small pockets of gas in them. These 'seed bubbles' form nucleation sites for prospective bubbles. If supersaturation is sufficient in the liquid then the dissolved gas is driven from the beer into the seed bubble; hence the reason that kegged beers are more fizzy. As gas is driven into the bubble it increases in size until it breaks away from the surface to which it was attached. At this point, the hydrostatic forces acting on the bubble overcome the surface tension forces that caused it to adhere to the surface. After breaking away, the bubble leaves a minute gas pocket behind, which represents the next seed bubble.

It is known that bubble formation is is directly related to the surface tension of a liquid; the lower the surface tension, the more bubbles produced. Ethanol, being a surfactant, reduces the surface tension of water, and beers up to about 5% ABV show increased foaming characteristics. Above this ethanol level there is, mysteriously, a decrease in the foamability of a beer!

The stabilty of beer foam depends on a number of factors, some of which appertain to any liquid, and some being unique to beer. In the latter category, probably the most important entities are the hydrophobic polypeptides which emanate from malt proteins, particularly hordein. These substances enter the bubble wall and stabilise it, thus decreasing the likelihood of collapse. Also significant are the isomerised hop resins, which are also hydrophobic and combine with polypeptides in the bubble wall. Strenghthening of the bubble wall causes the texture of each individual bubble, and hence the foam, to change; a sign of which is the phenomenon of lacing on the side of a glass whilst a beer is being drunk.

Although foams tend to form more readily at elevated temperatures, they are more stable at lower temperatures (because of the increase in viscosity of a liquid).

Excess foam in a beer is regarded as deleterious and is known as gushing, or 'wild beer'. Apart from the mis-handling of the product (*e.g.* over-agitating) a number of factors during production may stimulate gushing; one of these is over-carbonation. It is also known that using a barley grown under wet conditions and subsequently subjected to fungal spoilage by genera such as *Fusarium* and *Alternaria* also increases the likelihood of gushing.

Because N_2 is far less soluble in water than CO_2 the bubble size attainable with the gas is far smaller. This yields a far more compact and stable foam and is one of the main reasons for its increased use in the production of brewery-conditioned beers.

The capability of a beer to produce foam is relatively simple to assess and involves measurement of, for example, dissolved CO_2. Measurement of foam stability, however, is far more difficult, and numerous methods exist which vary from artificially inducing foaming in glass tubes (with the subsequent timing of re-formation of the liquid phase) to the use of conductivity probes. In the UK it is interesting to note that the IOB does not recommend any one method.

Even allowing for the enormous amount of research that has been carried out into the broad subject of head retention, there is still a mysterious side to the subject, and for an erudite appraisal of this facet of brewing the reader is referred to the article by Bamforth.[12]

NITROSAMINES

Since Piria demostrated the N-nitrosation of primary aliphatic amines in 1848,[13] there has been a wealth of literature published, not only on the nitrosation of this atom, but on other atoms (such as chlorine, carbon and oxygen) in a wide variety of compounds. For a comprehensive treatise, the reader is recommended to Williams[14] and, for the importance of N-nitrosation in the brewing industry, to Smith.[15]

Nitrosations have many industrial applications, being involved in the manufacture of rubber, azo-dyes and ε-caprolactam (nylon 6). In 1954, Barnes and Magee[16] demonstrated that certain diseases amongst rubber workers could be attributed to the exposure to N-nitrosodimethylamine (NDMA). Two years later, the same workers demonstrated that NDMA was carcinogenic to rats[17] and in 1979, Spiegelhalder et al.[18] demonstrated the presence of N-nitrosamines in beer, after they had only previously been reported from other foodstuffs. Over the past twenty years, N-nitrosamines (and their precursors), nitrate ions and nitrite ions have been the subject of much discussion in the food industry generally.

In simple terms, N-nitrosamines are formed by replacing a hydrogen atom that was attached to a nitrogen atom of an amide or amine by subjecting such compounds to a nitrosating agent. Thus, if diethylamine is nitrosated then NDMA results (Figure 6.4).

In the brewing industry the most significant nitrosating agents are oxides of nitrogen, i.e. NO_2, NO_3, N_2O_4 and $H_2NO_2^+$. The reactions leading to the formation of these intermediates are acid-catalysed and involve the addition of protons to nitrite ions, followed by reactions of nitrous acid. The amount of nitrosating intermediate present at any one time in the mixture is dependent upon pH; thus, in the range pH 2.0–5.0, all intermediates are detectable, whilst below pH 2.0 the nitrous acidium ion ($H_2NO_2^+$) predominates. The pH also affects the

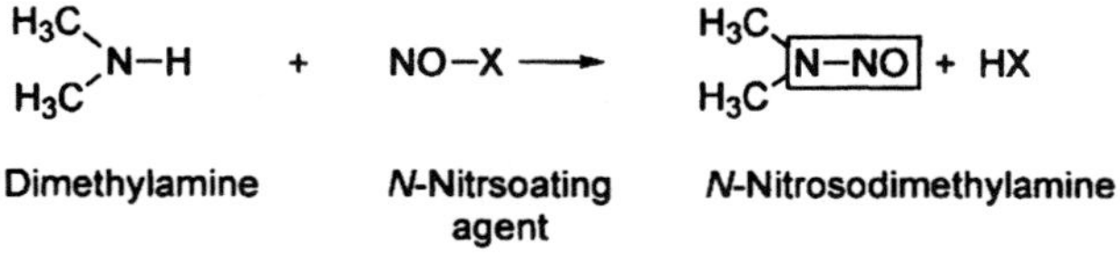

Dimethylamine   N-Nitrsoating   N-Nitrosodimethylamine
                          agent

Figure 6.4 *Formation of* N-*nitrosodimethylamine*

availability of amines for nitrosation; at too low a pH, amines are fully reduced and cannot be *N*-nitrosated.

Both volatile and non-volatile *N*-nitrosamines are found in beer, the former category (principally NDMA) originating from malt. Malt NDMA is produced during the kilning stage through a reaction between malt amines and oxides of nitrogen in the kiln gases. Since the detection in 1979 of *N*-nitrosamines in beer, malting companies have made considerable progress in reducing NDMA levels in malt, and a voluntary limit of 0.5 μg NDMA per kg of malt has been imposed. Malt NDMA levels are now so low that their contribution to *N*-nitrosamines in finished beer is negligible. Non-volatile *N*-nitrosamines are far more difficult to identify and quantify and, to overcome this problem, the Brewers' Society (now BLRA) and the BRF (now BRI) in 1987 developed a method for the measurement of 'apparent total *N*-nitroso compounds' (ATNC) in malt and beer.

It has now been established that most of the measureable ATNC in beer is caused by the microbial reduction of nitrates during the brewing process, principally during fermentation. Micro-organisms play an important role in the cycling of nitrogen in the biosphere, and nitrate-reducing organisms have been well documented. There are two main types:

(1) *Assimilatory nitrate-reducers.* These are capable of utilising the nitrate ion (NO_3^-) as the sole source of nitrogen for their cellular requirements. The uptake of nitrate is an energy-requiring process and is actively repressed by the presence of ammonia (or NH_4^+) or amino compounds in the wort. On uptake, nitrate is reduced to nitrite (NO_2^-) by nitrate reductase enzyme (NR), and the nitrite is reduced to ammonia by nitrite reductase (NiR). Both reductases are found in the cytoplasm of the organism concerned, and their production is induced by the presence of nitrate and nitrite, respectively, during aerobic growth. Again, their production is repressed by the presence of ammonia and/or amino nitrogen.

Certain species of wild yeast exhibit assimilatory nitrate reduction, principally *Brettanomyces anomalous, Hansenula anomala, Candida versatilis* and *Rhodotorula glutinis*, all of which have been isolated from cask-conditioned beers. Recent studies have suggested, however, that they are responsible for very little nitrate reduction *in vivo* and that they contribute little to the final ATNC of a beer.

(2) *Dissimilatory nitrate-reducers.* These are facultatively anaerobic organisms that undergo nitrate respiration in the absence of oxygen.

Under anaerobic conditions such organisms, some of which happen to be brewery contaminants, can use nitrate as an oxidant to remove electrons from the end of the respiratory chain. A typical electron donor in such circumstances is formate or lactate, with the reducing power being provided by enzymes such as formate dehydrogenase, *etc.* With the use of nitrate reductase (NR), nitrite is the subsequent end-product.

Under aerobic conditions these facultative organisms use oxygen as the terminal electron acceptor, with water being the end-product.

In contrast to assimilatory nitrate reduction, this dissimilatory mechanism is independent of wort amino-nitrogen concentration.

Certain dissimilatory nitrate reducers can produce nitric oxide (NO) and/or nitrous oxide (N_2O) *via* a metabolic pathway that involves a co-factor-dependent NiR.

Obesumbacterium proteus, a contaminant of yeast, possesses formate-dependent NR and NiR enzymes, which are extremely active under anaerobic conditions (oxygen concentrations below 5 mg l^{-1}). When pitched into wort (with yeast), the initial oxygen concentration is high enough to inhibit any nitrate reduction by *O. proteus*, but levels soon fall below the concentration mentioned above as fermentation commences and nitrite is produced, which is then chemically reduced to *N*-nitrosating species – with the subsequent production of NDMA. *O. proteus*-catalysed NDMA production occurs principally in the pH range 6–9, whilst chemically-catalysed production (from dimethylamine and nitrite) occurs most readily at around pH 3.5. Since wort pH falls during fermentation, it is likely that the chemically-catalysed route is going to be the major contributory factor to NDMA content in beer.

With the appreciation of its potential role in total ATNC production, modern hygiene regimes in breweries have led to a vast reduction in *O. proteus* populations, especially in pitching yeast. Even so, occasional high ATNC levels in beer can arise from breweries that have eradicated *O. proteus* and use malt known to be *N*-nitrosamine-free. The problem has been traced to certain thermophilic bacteria which proliferate as a result of practices such as wort-recycling, unhopped wort storage or faulty wort cooling. One such organism is *Bacillus coagulans*, which possesses both NR and NiR activity. In addition, this organism has the ability for further reduction of nitrite to nitrous oxide, ammonia, or even molecular nitrogen (*i.e.* it is a denitrifying organism). The possibility exists that there are some very powerful (as yet unknown) *N*-nitrosating agents produced as intermediates during these complicated reactions.

In 1987, the Brewers' Society (now BLRA) designated 20 µg (N-NO) kg^{-1} as a recommended upper ATNC limit in beer. In a survey conducted on 172 beers by MAFF in 1992 (Food Surveillance Paper No. 32), the average ATNC was found to be 54 µg (N-NO) kg^{-1}.

BEER FLAVOUR

The characteristics of a beer are contributed to by a mixture of colour, taste and smell, the last two being called organoleptic properties, which are obviously detected by the tongue and the nose respectively. The colour of a beer gives the drinker pre-conceived ideas, irrespective of overall flavour, *i.e.* if it is very light in colour then it must be a lager! For this reason, flavour assessment usually involves the participants wearing blindfolds or drinking from black-sided glasses.

Because of the nature and *raison d'être* of the product it must be remembered that there are no machines available for measuring beer flavour.

Beer flavour assessment, until relatively recently, had been a somewhat empirical exercise, whereby individual breweries would have a tasting panel competent to evaluate only their own products. In basic terms, there are three primary flavour groupings that have the highest impact on beer character. These are:

(1) bitterness – derived from α- and β-acids;
(2) mouthfeel – mainly derived from the degree of carbonation;
(3) alcohol – attributable to ethanol.

Even an experienced flavour assessor would be hard-pressed to distinguish and describe more than about 100 different beers.

It was not until the late 1970s that the American Society of Brewing Chemists (ASBC), the Master Brewers' Association of the Americas (MBAA) and the European Brewery Convention (EBC) masterminded a joint project aimed at adopting an internationally acceptable (and recognisable) set of beer flavour terms. As a result, the Beer

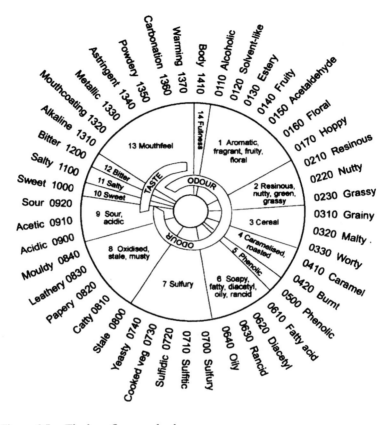

Figure 6.5   *The beer flavour wheel*
(Courtesy of the American Society of Brewing Chemists)[19]

Flavour Wheel was developed.[19] A simplified form is illustrated in Figure 6.5.

The wheel consists of 14 primary flavour classes, each divided into, so-called, first-tier terms; 44 terms in all. First-tier terms are all assigned a standard number and are further divided into a total of 78 second-tier descriptions (not shown in the figure above). Thus, the first-tier term 'fruity' (0140) is further divided into citrus (0141), apple (0142), banana (0143), blackcurrant (0144), melony (0145), pear (0146), raspberry (0147) and strawberry (0148).

Some sensations, such as bitterness, sweetness and saltiness are actually tasted, *i.e.* recognised by receptive areas on the tongue, but many so-called 'tastes' are, in fact, detected by nasal receptors and not the tongue. Hundreds of compounds contribute to the myriad of beer styles; even an 'ordinary' bitter or lager contains over 800 compounds capable of affecting flavour. Speciality beers, such as gueuzes, lambics and wheat beers contain even more flavour-active compounds – they are described further in Chapter 7.

REFERENCES

1 I. Clissold, *The CAMRA Guide to Cellarmanship*, CAMRA Books, St. Albans, 1997.
2 N.B. Redman, *The Brewer*, 1991, **77**, 106.
3 N.S. Fitch, *The Brewer*, 1998, **84**, 402.
4 G.G. Stewart, 'Proceedings of the 6th International Brewing Technology Conference', Harrogate, 1996, p. 182.
5 D.J. Cooper, G.G. Stewart and J.H. Bryce, *J. Inst. Brew.*, 1998, **104**, 83.
6 T. D'Amore, C.J. Panchal and G.G. Stewart, *J. Inst. Brew.*, 1987, **93**, 472.
7 T. D'Amore, G. Celotto and G.G. Stewart, 'Proceedings of the 23rd European Brewing Convention', Lisbon, 1991, p. 337.
8 F. Schur, 'Proceedings of the 19th European Brewing Congress', London, 1983, p. 353.
9 C.W. Bamforth, *Beer*, Plenum Press, New York, 1998.
10 C.R. Murray and W. Van der Meer, *US Patent* 5 304 384, 1994.
11 Institute of Brewing, *Methods of Analysis, Vol. 1 – Analysis*, January 1997.
12 C.W. Bamforth, *The Brewer*, 1995, **81**, 396.
13 R. Piria, *Annalen*, 1848, **68**, 343.
14 D.L.H. Williams, *Nitrosation*, Cambridge University Press, Cambridge, UK, 1988.
15 N.A. Smith, *J. Inst. Brew.*, 1994, **100**, 347.
16 J.M. Barnes and P.N. Magee, *Br. J. Ind. Med.*, 1954, **11**, 167.
17 P.N. Magee and J.M. Barnes, *Br. J. Cancer*, 1956, **10**, 114.
18 B. Spiegelhalder, G. Eisenbrand and R. Preussmann, *Food Cosmetics Toxicol.*, 1979, **17**, 29.
19 M.C. Meilgaard, C.E. Dalgleish and J.F. Clapperton, *J. Am. Soc. Brew. Chem.*, 1979, **37**, 47.

Chapter 7

Microbiology in the Brewery

All of the major raw materials used for brewing are potential sources of unwanted micro-organisms, none of which (unless there is a sudden water-borne outbreak of disease) are likely to be pathogenic to man. In addition, brewing aids such as finings, primings and filtration media, and containers (casks, bottles, *etc.*) can contribute contaminants.

Both wort and beer are prone to spoilage organisms, the former especially so since it provides a nutrient-rich, oxygenated environment. In theory, many micro-organisms can grow in wort, particularly if its temperature is allowed to fall during processing. Good brewing practice should ensure that this does not happen and worts from the mash tun should, ideally, pass straight to the copper for boiling. Beer itself, with its intrinsically low pH, does not provide a hospitable environment for the survival of most bacteria. It is also largely devoid of the nutrients required for microbial growth, particularly carbon sources; these have, of course, been consumed by yeast during fermentation. In addition, ethanol has an inhibitory effect on most microbes and the hop iso-α-acids are known to possess antimicrobial activity, although the actual mode of action is, as yet, unknown. Packaged beer, especially in bottle, can and keg, should be totally devoid of oxygen and only capable of supporting the growth of a few anaerobic microbes. No known pathogenic anaerobes can survive in beer, although, as we shall see, one or two anaerobic bacteria can be problematical in packaged products.

Microbes emanating indigenously from brewing liquor, malt and hops, and those found in wort, will not survive the boiling stage, and this, together with the fact that the use of potable water is statutory in the food industry, means that there are relatively few species of bacteria and fungi that cause spoilage problems; the list is restricted to a few wild yeasts and certain Gram-positive and Gram-negative bacteria. Wild yeasts, by definition, are those *Saccharomyces* and non-*Saccharomyces* spp. that are not the strains specified for fermentation. There are no reported instances of any human pathogen being contracted through drinking beer!

Some micro-organisms produce highly resistant stages called spores. Resistance is usually to temperature change, pH, chemical agents, lack of oxygen and starvation. Certain spores can survive the various processing stages in the brewery and end up in packaged beer where they 'exist', *i.e.* they may not necessarily germinate and re-liberate their vegetative phase.

By applying the Hazard Analysis Critical Control Point (HACCP) concept, the Institute of Brewing has produced a generally-accepted scheme for microbiological process control within the brewery (Figure 7.1). Microbiological and certain physical (*e.g.* temperature) analyses should be carried out according to recommended Institute methods[1,2] and at each stage in the brewing cycle where quality can be critically affected. Having identified critical control points and relevant sampling points, the scheme can only operate efficiently if results can be obtained within a realistic time period. Traditional microbiological methods[3] involve aseptic sampling (*e.g.* swabbing) and subsequent encouragement of growth of organisms contained in the sample. For identification purposes this invariably necessitates the production of visible colonies of the organism concerned on a solid growth medium (agar plate). Depending upon the microbe being monitored, results are rarely forthcoming until a 24 hour incubation period has elapsed; sometimes a 3–4 day incubation is necessary. Thus, with such methodology, it is only possible to assess a process, such as cask-washing, for microbiological sterility rather than to monitor it; results can only be used in the context of hindsight.

With modern, automated, cleaning-in-place (CIP) systems it is no longer satisfactory to wait for results; one must know whether the cleaning regime is actually working. Accordingly, there are now some rapid methods available which enable results of basic microbiological analyses to be obtained within a few hours, or even minutes. This enables the operator to alter the cleaning regime straight away if necessary. Rapid methods readily available at present do not really allow specific determination of spoilage organisms to be performed in a very short time.

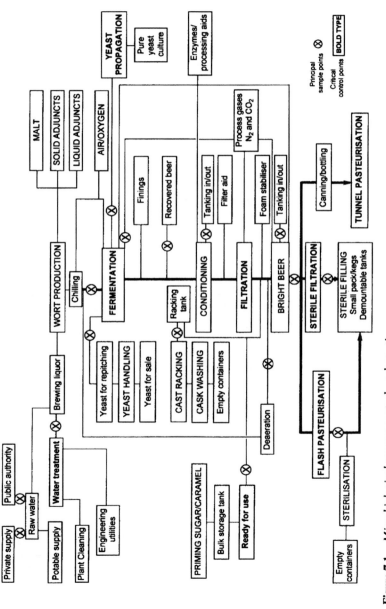

Figure 7.1 *Microbiological process control – schematic*
(Courtesy of the Institute of Brewing[2])

ATP BIOLUMINESCENCE

The universal presence of ATP in living organisms and the reaction of this adenosine compound, with the substrate luciferin and the enzyme luciferase with the subsequent emission of visible light (562 nm), has provided the food microbiologist with a means of rapidly assessing the microbial status of his/her plant. Many foodstuffs, including beer, naturally contain ATP as a biological residue and this has to be allowed for when interpreting results.

The biochemical reaction (7.1) employed is that which causes the tails of fireflies to glow and requires the presence of oxygen and Mg^{2+},

$$ATP + O_2 + \text{D-luciferin} + Mg^{2+} \xrightarrow{\text{luciferase}} AMP + CO_2 + \text{oxyluciferin}$$
$$+ PPi + \text{☼} \ (562 \text{ nm}) \quad (7.1)$$

where AMP is adenosine monophosphate, and PPi is inorganic pyrophosphate.

The amount of light produced is measured using a luminometer and is directly proportional to the amount of ATP present (*i.e.* ATP $\cong$ number of organisms $\cong$ contamination). ATP presence can also be due to dead organic material (*e.g.* soil) as well as viable organisms, and so the bioluminescence technique can also evaluate a general lack of hygiene.

The test, in its simplest form, involves taking a swab sample from the suspect site and introducing it directly into the luminometer, into a chamber which contains measured amounts of ATP extractant, luciferin, luciferase and other reagents necessary for ATP to instigate the 'firefly' reaction. Modern machines are portable and highly sensitive. The 'lightning' system developed by Idexx Laboratories, Westbrook, Maine, USA also incorporates agents which neutralise compounds known to cause interference to the reaction and, hence, lead to erroneous results. Their system is also insensitive to detergents which have also caused interpretation problems in some luminometers, although it has the advantage that results can be obtained within one minute.

Measurement of ATP solely from viable micro-organisms from a sample point takes longer because separation processes are involved. The normal procedure here is to take rinse samples and subject them to membrane filtration. The filters are washed with ATPase to remove extracellular ATP. Intracellular ATP is then extracted from the cells on the membrane and assayed by the luciferin–luciferase method. The initial wash and ATPase treatment mean that the results can take about one hour to be obtained, which is still realistic enough to enable remedial action to be taken. By incubating the membrane on a selective growth medium, and then washing and assaying, one can also monitor the numbers of specific groups of microbes; individual species or those growing in very small numbers (depending on how selective the medium is). The incubation period, which would be around 24 hours, obviously lengthens the period required between sampling and production of results. According to Hammond,[4] the correlation between ATP bioluminescence results and those obtained from traditional microbiological methods is very good.

THE POLYMERASE CHAIN REACTION (PCR)

Modern recombinant DNA techniques have led to a method of detecting contaminating micro-organisms even though they are only present in small numbers. Results can be obtained in a shorter time period than those obtained from traditional microbiological methods, although not as rapidly as those from ATP bioluminescence methods. The technique involves amplifying very small fragments of DNA, from the 'foreign' microbe, which can then be made visible by gel electrophoresis or by an enzyme-linked immunosorbent assay (ELISA). Because the technique is very sensitive, extreme care must be taken when handling samples. Amplification of DNA is by means of primers, and so even DNA from dead cells will be detected. Early attempts to use this method for identifying beer contaminants met with some difficulties, principally because of interference from yeast itself, or from various beer constituents. Many of these problems have now been overcome. The specificity of PCR is attributable to the nature of the nucleotide primers used during the amplification process. Primers can be broad spectrum (*i.e.* for bacteria in general) or narrow spectrum (for individual genera or species). The latter approach is normally referred to as specific PCR.

An enormous amount of research has been conducted since the mid-1990s and adaptations of the original PCR concept have led to some remarkably specific identifications to be effected. Amongst several potentially useful quality-control methods to emanate from PCR-based work in the last few years, mention may be made of the following:

(1) *Detection of lactic acid bacteria in yeast cultures.*[5] This method involves the use of a technique called 'nested' PCR, whereby an

initial PCR is carried out using a broad spectrum primer, which is then followed by a second PCR on the amplified product of the first. The primers used in the second stage bind exclusively to lactic acid bacteria and are specific to the genera *Lactobacillus, Pediococcus* and certain species of *Leuconostoc.* Use of these more specific primers in the first stage proved impossible because of interference from yeast cells and certain beer constituents, but this nested method permits very low levels of lactics to be detected in vast numbers of yeast cells (*i.e.* fermentation conditions). Total detection time is approximately eight hours.

(2) *Differentiation of ale, lager and wild (non-brewing) yeasts of the species* Saccharomyces cerevisiae.[6] Again, a double series of PCRs is employed, the first being designed to distinguish ale and lager strains from wild strains and bacteria. The second PCR involves the use of random amplified polymorphic DNA (RAPD), which employs very short and specific nucleotide primers. This step enables different varieties of *Sacch. cerevisiae* to be defined within the same culture. It was noticed that the RAPD-PCR primer has a higher affinity for the ale strain of the yeast than it does for the lager and non-brewing strains.

(3) *General microbiological analysis of beer.*[7] This very sensitive method, again employing nested PCR, was developed at the Technical University of Munich, Weihenstephan. The intial PCR employs a broad spectrum primer for bacteria generally. This is followed by a series of specific PCRs with primers for *Lactobacillus* (six spp.), *Pediococcus* (two spp.), *Pectinatus* and *Megasphaera.* The documented method, as it stands, was deemed to be too lengthy for use in routine analyses but it can detect 10^2–10^3 bacterial cells in a yeast suspension comprising 20×10^6 cells.

A scheme for the detection and enumeration of micro-organisms important in brewing is shown in Figure 7.2.

BACTERIA

True bacteria (Eubacteria) are placed into two main groups according to whether they take up Gram's stain (gentian violet) irreversibly (Gram-positive) or reversibly (Gram-negative). For reversible uptake, gentian violet can be removed by a decolourising agent (acetone or ethanol) and the cells counter-stained with a red dye. Thus organisms

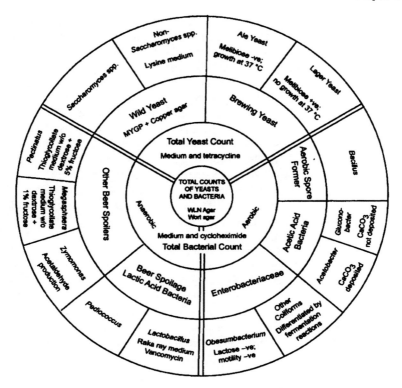

Figure 7.2   *Scheme for the detection and enumeration of micro-organisms*
important in brewing
(Courtesy of the Institute of Brewing[2])

are classified as Gram-positive (cells appear blue, or violet) or Gram-negative (cells appear red). A few bacteria are Gram-variable, and some Gram-positive bacteria become Gram-negative with age, but the stain is still used as a fundamental step for bacterial diagnosis.

Most of the bacteria likely to be encountered in the brewery exist in either a cigar-shape (called a rod, or bacillus) or as a sphere, or coccus. Cocci can occur singly or in packets and chains. Gram-negative bacteria have a different, more complicated cell wall structure than Gram-positives and are generally more resistant to chemical agents (including antibiotics).

The Institute of Brewing scheme for the identification of Gram-positive bacteria is shown in Figure 7.3 and that for Gram-negative bacteria is illustrated in Figure 7.4.

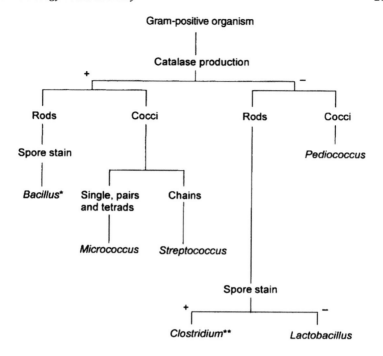

Bacilli usually stain Gram-positive in early stages of growth

**Clostridia* usually stain Gram-positive at least in very early stages of growth, although in some species Gram-positive cells have not been seen.

Figure 7.3 *Scheme for the identification of Gram-positive bacteria* (Courtesy of the Institute of Brewing[2])

Enterobacteriaceae

One of the larger groups of Gram-negative bacteria are the Enterobacteria (family, Enterobacteriaceae). These exist as short, non-spore-forming rods which may be flagellated (motile) or non-flagellated (non-motile). Flagella, when present, are situated around the periphery of the cell (peritrichous). Bacteria in this class have the ability to grow in the presence of bile salts, which are inhibitory to the majority of organisms; indeed, most are 'gut organisms', being a natural part of the gut flora of humans and other animals. This sort of habitat necessitates their being able to grow in the presence or absence of oxygen and they are, in fact, remarkably tolerant to changes in

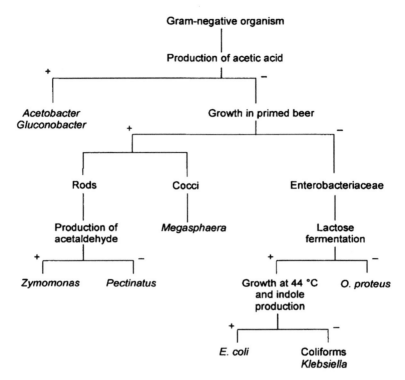

Figure 7.4   *Scheme for the identification of Gram-negative bacteria*
(Courtesy of the Institute of Brewing[2])

oxygen tension. They are referred to as being facultatively anaerobic bacteria. The most celebrated example is probably *Escherichia coli*, a much-studied organism because of its short generation time and ability to grow readily on synthetic media. This, and related organisms, comprise the coliforms.

Enterobacteriaceae are differentiated on the basis of biochemical (metabolic) characters, *e.g.* the ability to ferment a range of sugars, the ability to use citrate as a sole source of carbon, the production of indole, and the production of acetoin (acetyl methyl carbinol), *etc.* Most can reduce nitrates to nitrites, most produce catalase, and none produce oxidase. Fortunately for the brewer they are not tolerant of alcohol and cannot grow in a medium with a pH of less than 4.3; so they cause no problems in finished beer, even though one or two can survive in it and have been isolated from it. They can, however, be troublesome organisms in wort, both before and during fermentation.

Citrobacter freundii and *Enterobacter agglomerans*, known from foodstuffs generally, have been isolated from brewery plant and from wort, but the most important organism in this group is *Obesumbacterium proteus*.

Obesumbacterium proteus

The 'short, fat rods' described late last century as being a common occurrence in pitching yeast and in fermentations have had a very checkered history in terms of their nomenclature. These seemingly ubiquitous bacteria were first named as being *Flavobacterium proteus* by Shimwell and Grimes in 1936.[8] The specific epithet 'proteus' was coined because the cells were very variable in shape (pleomorphic) especially when grown under different conditions. Proteus was the ancient Greek sea god with the ability to change shape at will. A pure culture was isolated from the Beamish and Crawford brewery in Cork, but subsequently became mislaid. Even the culture deposited in a stock culture bank disappeared! With the absence of a viable sample it was impossible to equate Shimwell and Grimes's original descriptions and characteristics of their strain with modern bacteriological nomenclatural requirements, and it was not until 1955 that a type strain from the Shaefer Brewing Co., Brooklyn, NY, was deposited in a culture collection (ATCC 12841) by Strandskov and Bockelmann.[9]

As taxonomy progressed, *Flavobacterium* spp. were re-defined as being oxidative organisms, and so *F. proteus* was removed because it ferments D-glucose and other carbohydrates. Subsequent DNA/rRNA hybridisation experiments on '*F. proteus*' confirmed that it was not related to the genus *Flavobacterium* at all.

In 1963, Shimwell[10] propounded a new genus for the 'short, fat rods', the genus being *Obesumbacterium* and the species being *O. proteus*. On the basis of further DNA/RNA hybridisations on several strains, Priest *et al.*[11] suggested that *O. proteus* should be transferred to the genus *Hafnia* (Hafnia being the old name for Copenhagen) and recommended re-naming it to *H. protea*. This was accepted for a while, but *H. protea* disappeared in 1980 (except in brewing literature) because of lack of standing in official international nomenclatural lists (*O. proteus* was so listed and, therefore, accepted by taxonomists). Subsequent work (Brenner[12]) showed that there are two biogroups of the organism, 1 and 2, and that *O. proteus* biogroup 1 is synonymous with *Hafnia alvei* (the type species), which has become adapted to the brewery environment. It was recommended that *O. proteus* should be regarded as 'two separate species' (sic) with the type strain having the properties of biogroup 1.

O. proteus can tolerate the harsh environment of a vigorous fermentation and actually binds to yeast cells, on the surface of which its numbers become concentrated. On solid culture media it is slow-growing and there is little visible sign of growth until 48 hours have elapsed. When present in significant numbers during fermentation it can produce elevated levels of diacetyl, dimethyl sulfide and fuselols in the resultant beers. *O. proteus* is more frequently found in top-fermentations and is relatively rare in lager breweries.

Zymomonas mobilis

This rod-shaped, Gram-negative bacterium is unique in that it can quantitatively ferment glucose, fructose and sucrose (some strains) and produce equimolar amounts of ethanol and CO_2. This is effected by the Entner–Doudoroff (or keto-deoxygluconate) pathway (Figure 7.5); trace amounts of lactic acid are produced as well. In these organisms, 6-phosphogluconate is dehydrated and pyruvic acid and glyceraldehyde-3-phosphate are produced. From these two C_3 intermediates, ethanol is liberated by reactions of the EMP pathway. It also differs from the Enterobacteriaceae because of its inability to reduce nitrates, its alcohol tolerance and its polar flagella (situated at the ends of the cell). It can also grow in media with a pH of less than 4.3 – as low as 3.5. In older literature the organism has appeared as *Saccharomonas lindneri* or *Pseudomonas lindneri* after P. Lindner, who first described the bacterium.

It was originally isolated from fermenting palm juice and has also been recovered from sweet cider where it causes 'cider sickness', which is, in essence, a secondary fermentation (much frothing) resulting in loss of sweetness and with taste and odour changes. In the brewing industry, *Z. mobilis* has only been isolated from cask-conditioned beers, principally those that have had priming sugars added to them. It is unlikely to cause wort infection since it is unable to ferment maltose. The brewery form is now regarded as *Z. mobilis* sub-spp. *mobilis*, whilst the cider contaminant is *Z. mobilis* sub-spp. *pomacii*. Fortunately, the bacterium is rare because spoilage is rapid once infection has occurred. Beers are more prone to infection during warmer weather, especially if there is some failure in the cask-washing regime. The main signs of spoilage are turbidity and off-odours due to hydrogen sulfide and acetaldehyde.

Zymomonas could find considerable usage as an industrial ethanol

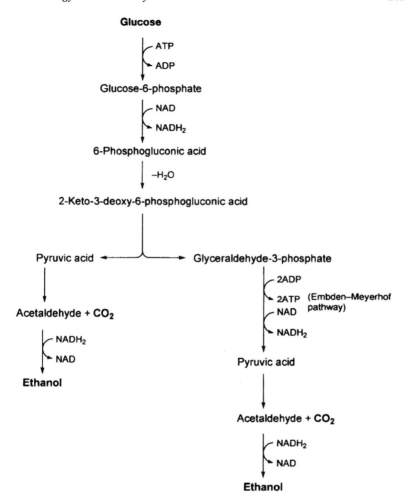

Figure 7.5 *The Entner–Doudoroff pathway*

producer; it exhibits higher rates of glucose uptake and ethanol production than *Saccharomyces cerevisiae* and does not yield such large quantities of unwanted biomass for disposal.

According to Swings and DeLey,[13] *Zymomonas* is genetically, phenotypically and ecologically related to the acetic acid bacteria (see next section); both types of bacteria occur in acid, sugary and alcoholised habitats. They propose that both types of organism could act in a synergistic way in certain natural habitats; *Zymomonas* producing alcohol and acetic acid bacteria oxidising it. *Zymomonas* is closer to *Gluconobacter* because of its polar flagella, incomplete tricarboxylic acid cycle and Entner–Doudoroff pathway. There is one slight problem, however, as we shall see; acetic acid bacteria are strictly aerobic.

Acetic Acid Bacteria

These Gram-negative, aerobic rods have the propensity to oxidise ethanol and produce acetic acid. There are two main genera, *Acetobacter* and *Gluconobacter*, which are closely related and grouped together in the family Acetobacteriaceae. Both genera are alcohol- and acid-tolerant and their use of alcohol as a carbon source is *via* an oxidative pathway. The main distinguishing features between the two genera are that *Acetobacter* spp. have peritrichous flagella (if motile) and can oxidise acetic acid and lactate to CO_2, whereas *Gluconobacter* spp. have polar flagella (if motile) and cannot oxidise acetic acid and lactate to CO_2. In the early days of brewery microbiology these were known as the 'strong' and 'weak' oxidisers of ethanol, respectively, and were placed in the same genus (*Acetobacter*).

The most likely species of *Acetobacter* to be encountered in the brewery is *A. pasteurianus*, which is a particular problem in cask-conditioned beers, especially at the point of dispense. When beer is drawn from a cask, the air replacing it invariably contains some acetic acid bacteria. There is even some evidence that the organisms are spread by fruit flies in warmer weather. As well as using ethanol as a carbon source, *Acetobacter* spp. can also utilise sugars *via* the hexose monophosphate pathway and tricarboxylic acid cycle. Although fundamentally aerobic, these organisms can grow under conditions of very low oxygen tensions (microaerophilic); even so, they rarely appear to be contaminants of keg beers or other packaged beers held under gas.

Most authorities agree that there is only one species of *Gluconobacter*, *G. oxydans*, although there are a number of sub-species. *G. oxydans* is obligately aerobic with a purely respiratory mode of metabolism. Its growth in beer often leads to a surface raft of growth, called a pellicle, and eventually general turbidity. Some strains produce copious amounts of dextrans and levans which increase the viscosity of the infected beer and cause a condition called 'ropiness'. As well as having a partiality for ethanol, *G. oxydans* thrives on a wide variety of sugar substrates, especially soft fruits where severe spoilage can occur. The organism has also been found in soft drinks.

Growth experiments in defined media show that it can utilise many carbon sources, with preference being shown for D-mannitol. The order of preference for carbon seems to be: D-mannitol, sorbitol, glycerol, D-fructose, D-glucose. The hexose monophosphate pathway enzymes are all present, and this route represents the only means of complete breakdown of sugars and polyols. The requirements for the tricarboxylic acid and glycolytic pathways are not complete and partial breakdown products result. This makes *Gluconobacter* a potentially useful industrial organism and under the appropriate conditions one can manufacture: dihydroxyacetone from glycerol; L-sorbose from D-sorbitol; D-tartaric acid, D-gluconic acid, 5-ketogluconic acid and 2-ketogluconic acid from D-glucose.

In brewing literature, there appears to have been some taxonomic confusion in yesteryear and *Acetomonas*, a not generally accepted genus, seems to have been analagous with *Gluconobacter*.

Pectinatus cerevisiiphilus

This Gram-negative, non-spore-forming rod was first isolated in the USA, in 1971, from beer which had been held in a ruh storage tank. Originally, it was provisionally identified as a species of *Zymomonas* until it was realised that: it did not produce ethanol or acetaldehyde from glucose; it did not have polar flagella (flagella all emanated from one side of the cell); and that it was obligately anaerobic. The rod is distinctly curved and the flagella are all situated on the concave wall which gives the organism a comb-like appearance (hence the generic name). After subsequent biochemical work, Lee *et al.*[14] proposed a new genus, *Pectinatus* (family, Bacteroidaceae), and within a few years the bacterium had been recorded from Germany, Finland, Japan and France – and fully-named *P. cerevisiiphilus*. It is now evident that there are at least three different strains implicated in beer spoilage. Once a beer becomes infected its deterioration is rapid due to the production of metabolic by-products such as propionic acid, acetic acid, succinic acid and, to a lesser extent, lactic acid. Hydrogen sulfide is also released in considerable quantities and this, in combination with non-volatile and volatile fatty acids, produces a nasty 'rotten eggs' smell. These odour defects manifest themselves long before the infected product assumes turbidity, which can take at least five days.

Fortunately, the bacterium is sensitive to heat (one minute at 58 °C being lethal – which is less than one pasteurisation unit). Cells are also sensitive to normal brewery sanitisers such as chlorine-releasing agents.

Megasphaera cerevisiae

This contaminant Gram-negative coccus (cocci occurring singly) was originally isolated and described by Weiss *et al.* in 1979[15] from a bottled German pilsner. It had even more strict anaerobic growth requirements than *Pectinatus*. Unlike *Pectinatus*, the genus *Megasphaera* already existed, the type species being *M. elsdenii* (family, Veillonellaceae) a natural inhabitant of the mammalian gut and rumen. DNA/DNA hybridisation indicated that this organism was not compatible with the brewery isolate, and in 1985 the name *M. cerevisiae* was proposed for the new species by Engelmann and Weiss.[16]

Growth experiments with defined media (peptone-yeast extract-based) indicated that glucose and maltose were utilised very poorly by *M. cerevisiae* and that pyruvate and lactate were far more preferable as carbon sources. Fructose is also well utilised. Growth can occur between 15 and 37 °C, the optimum being 28 °C. Biochemically, *M. cerevisiae* proved to be catalase negative, did not produce indole and could not reduce nitrates. Hydrogen sulfide is always produced, and this, together with a range of metabolic waste products, contributes to very unpleasant odours in spoiled beer. Typical metabolites include the organic acids: acetic, propionic, iso-butyric, *n*-butyric, iso-valeric, *n*-valeric and caproic.

The bacterium seems more prone to spoil low alcohol beers, especially those that have not been pasteurised. Its tolerance to alcohol seems to be about 2.8% ABV. *M. cerevisiae* has also been isolated from wort, in which it can grow under strict anaerobic conditions.

One of the reasons why *Pectinatus* and *Megasphaera* are fairly recent introductions to the list of beer spoilage organisms is that brewery microbiologists never really paid much attention to strict anaerobes as potential spoilers. This was because there were no stages in beer production that provided a purely anaerobic environment (*i.e.* they were never looked for). It is only with the recent introduction of filling apparatus, capable of producing vastly reduced levels of oxygen in a packaged product, that these bacteria have manifested themselves as spoilage organisms. It should be noted that *Pectinatus* and *Megasphaera* have only been recovered from large, modern breweries with up-to-date packaging facilities.

Because of their fastidious nature, particularly with respect to oxygen, they should be easier to contain than other beer spoilage bacteria. Most brewery sanitisers, especially chlorine-releasing agents, seem to be effective in their eradication.

Lactic Acid Bacteria

These are Gram-positive bacteria that actively spoil beer and must always be regarded as potentially dangerous since they can grow vigourously in situations of low oxygen tension (microaerophilic) and are also tolerant of alcohol and low pH. The lactics are less sensitive to hop iso-α-acids than the majority of Gram-positive bacteria and there are indications that strains indigenous to breweries gradually develop resistance to hop compounds. There are two principal genera implicated in beer spoilage, *Lactobacillus* and *Pediococcus*, and these are distinguished primarily on morphology. *Lactobacillus* spp. are rod-shaped and non-motile, whilst pediococci are spherical, the cocci often occurring in packets of four (tetrads), although they can occur singly or in pairs.

They are all non-spore-forming, do not produce catalase and cannot reduce nitrate. Their growth requirements are complicated and nutritionally they are very fastidious. Most need to have amino acids, vitamins and various growth factors supplied in their growth media, be it in artificial media or in nature. This is seen to be a sign of their high degree of adaptation to a particular environment. All have the ability to ferment a variety of sugars strongly (*via* a number of pathways) with the production of lactic acid. The degree of lactic acid production varies; in some species it is the major end product – these are called homofermentative lactic acid bacteria. Other strains yield a variety of other fermentation end-products, such as ethanol and acetic acid, and these are designated heterofermentative.

Homofermenters convert glucose (say) to pyruvate *via* the EMP pathway and then reduce pyruvate to lactic acid *via* a dehydrogenase (really a pyruvate reductase); pyruvate acts as a hydrogen acceptor here. Conversion of glucose can be almost quantitative. Heterofermenters are devoid of two essential glycolytic pathway enzymes: aldolase and hexose isomerase and dissimilate hexoses and other sugars *via* the phosphoketolase pathway [or hexose monophosphate (HMP) shunt] (Figure 7.6). In this type of fermentation the phosphorylation of glucose is immediately followed by an oxidation to 6-phosphogluconic acid, this compound being subsequently oxidised to CO_2 and pentose phosphate. The latter then yields the C_3 and C_2 intermediates glyceraldehyde-3-phosphate and acetyl phosphate, respectively. Ethanol is then produced from acetyl phosphate by two successive reductions (both involving $NADH_2$), and glyceraldehyde-3-phosphate yields lactic acid *via* reactions of the EMP pathway. Acetyl phosphate is formed from a pentose phosphate *via* a mechanism

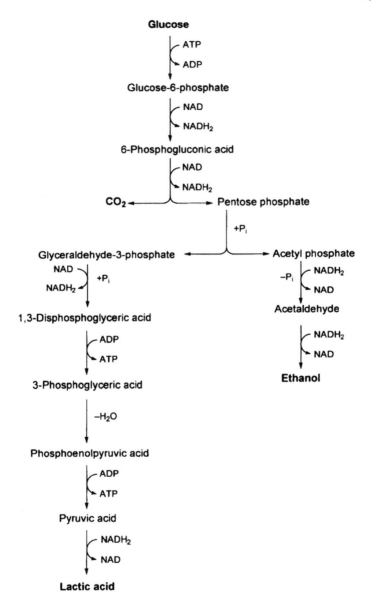

Figure 7.6 *The hexose monophosphate (phosphoketolase) pathway*

involving phosphoketolase and thiamine pyrophosphate, the C-1,2 fragment being cleaved and then rearranged (inorganic phosphate being taken up).

The net yield of ATP from glucose *via* HMP is 50% of that yielded by EMP because the energy-rich bond of acetyl phosphate is lost during its reduction to ethanol. Figure 7.6 shows that it is the breakdown (oxidative decarboxylation) of 6-phosphogluconate, into CO_2 and pentose phosphate, that determines the molar equivalence of the three end-products of a lactic heterofermentation.

Of the homofermenters likely to cause problems in the brewery, *L. casei, L. plantarum* and *L. delbrueckii* are most frequently encountered, the latter being thermophilic (optimum growth at 45 °C). *L. casei* and *L. ·plantarum* are closely related, both showing optimum growth at 30 °C. Principal differences between the two are their sugar fermentation profiles, the optical status of lactic acid produced and their natural habitats. *L. casei* is found naturally in milk and milk products where it plays an important role in cheese manufacture. This species can produce lactic acid (predominantly the D-form) from maltose, lactose, glucose, fructose, mannose, galactose and mannitol. *L. plantarum* is part of the natural surface flora of many plants and is used commercially in the manufacture of sauerkraut. It produces optically inactive lactic acid from maltose, sucrose, lactose, glucose, fructose, galactose, arabinose and raffinose. *L. delbrueckii* has been recovered from pre-cooled wort held at 60 °C, and quickly grows on spent grains if they are not disposed of straight away. L-Lactic acid is produced from maltose, sucrose, glucose, fructose, galactose and dextrin. This thermophile is introduced into the mash of certain beers, particularly lambics (see pages 217–222) and *Kaffir*, in order to lower the pH to about 3.0–3.5.

Heterofermentative lactobacilli are far more prevalent in breweries, with *L. brevis* being by far the commonest beer spoilage bacterium (certainly in German breweries). It is also known from milk, cheese and spoiled acidic foodstuffs. It shows optimum growth at 30 °C and produces optically inactive lactic acid from maltose, sucrose, glucose, fructose, galactose, xylose and arabinose. The closely related *L. pastorianus* is also the cause of souring of beer and has been isolated from distilleries as well. It too favours 30 °C and produces acid from maltose, sucrose, glucose, fructose, galactose, arabinose, raffinose, trehalose, dextrin and mannitol. Some strains of both of these species can produce slime (glucan) from sucrose.

Of the *Lactobacillus* species mentioned above as being recorded from the brewery environment, only a few strains within each species can actually tolerate the said conditions. The situation in *L. lindneri* is different, however, because all known strains can cause beer spoilage. *L. lindneri* was originally described and named earlier this century as being either *Bacillus lindneri* or *Bacterium lindneri*. Later on, microbiologists deemed it to be a variety of *L. brevis*; Eschenbecher in 1968[17] designating it *L. brevis* var. *lindneri*. In 1996, Back *et al.*,[18] on the basis of DNA hybridisation experiments, proved that it was a species in its own right. It has been isolated from a number of different sites within the brewery and it is estimated that *L. lindneri* was responsible for 25% of the microbiological problems encountered in German breweries during the period 1980–1990. The bacterium is highly insensitive to hop iso-α-acids and thermotolerant. In the latter context it can tolerate pasteurisation up to 17 PU. All other lactobacilli and pediococci fail to survive 15 PU. There is also some resistance to certain brewery sanitisers. Add to this the fact it can grow very rapidly in beer, and that it periodically produces unusually small cells which can pass through beer filters (it has been isolated from bright beer tanks), and *L. lindneri* assumes the status of being a very problematical microbe. In laboratory culture, growth is very slow, making assessment difficult, and occurs between 15 and 45 °C.

The genus *Pediococcus* has, in the past, been confused with *Sarcina* (sarcina = latin for 'package') which is another Gram-positive bacterium that typically has cocci in packets of eight and in older brewery literature, beer was said to have contracted 'sarcina sickness' if Gram-positive cocci were identified in spoilt batches. The organism is now called *Pediococcus damnosus* (it was *P. cerevisiae*) and it is generally far more problematical than lactobacilli. *P. perniciosus* has even appeared in the literature from time to time, but has never been accepted by taxonomists.

Pediococci cause turbidity, acid notes and unwanted flavours, and favour growth at lower temperatures than lactobacilli. Optimum growth occurs at 25 °C but they can tolerate 7 °C, hence their prevalence in lager breweries. Some strains can cause ropiness by exuding large quantities of viscous capsular material.

Diacetyl is produced during growth, but intracellularly rather than by the extracellular oxidation of the precursor, α-acetolactate, that occurs with yeast. This means that the mechanism of synthesis in those lactic acid bacteria capable of producing diacetyl is clearly different from that encountered in *Saccharomyces cerevisiae*.

The key intermediate, pyruvic acid, is metabolised by all known strains of lactics under anaerobic conditions, with the release of a wide variety of end-products: D- and L-lactic acid, ethanol, CO_2, acetoin (acetyl methyl carbinol), 2,3-butanediol and acetic acid. Pyruvate-reducing systems are far more active than pyruvate-oxidising ones. It was originally thought that diacetyl was produced from pyruvate *via* acetoin and the butanediol pathway, but evidence now suggests that α-acetolactate (AAL) is the precursor. This is certainly true for *L. casei*. The conversion of AAL to diacetyl can occur apparently both *via* enzyme mediation and non-enzymatically, it being irreversible. The reduction products of diacetyl, acetoin and 2,3-butanediol, however, can interconvert. As a group, lactic acid bacteria can utilise a number of primary carbon sources for diacetyl production, including malic acid, citrate and a variety of sugars.

Diacetyl is usually only produced in trace amounts by *Lactobacillus* spp., particularly in those likely to be found in beer; levels of 0.1 ppm are rarely exceeded. Pediococci, on the other hand, produce the compound in greater quantities and levels of 3 ppm can be attained; such levels are totally unacceptable for most beers. It should be noted that in the manufacture of wine and cheese, diacetyl can be a cherished part of the overall taste spectrum, up to 4 ppm being essential to the flavour of certain styles of wine.

Bacillus spp.

These are Gram-positive, endospore-forming, nitrate-reducing rods that can occur in wort from time-to-time. Multiplication there produces acidic metabolic waste products and a rise in ATNC levels. Spores can occasionally survive the boil and enter the fermentation stage and even the beer itself. Spoilage rarely ensues because, even if the spores germinate, the released vegetative cells will not survive in the conditions in which they find themselves.

WILD YEASTS

As already stated, wild yeasts are those that are present during fermentation and finished beer, but which are not wanted by the brewer; thus, top-fermenting strains (*Sacch. cerevisiae*) that occur in a lager fermentation (employing *Sacch. uvarum*) are considered to be wild, and *vice versa*. The overall definition encompasses a wide range of fungi, the nomenclature of which has always presented taxonomists with many a dilemma. Fungi that produce sexual, or perfect, spores (*i.e.* the result of a mating process) can be fairly readily classified into three major groups: Phycomycetes, Ascomycetes and Basidiomycetes, according to the types of spore produced. Those fungi that do not undergo sexual spore formation, by definition, are classified as imperfect – Fungi Imperfecti. Mycological studies over the years have determined that many genera, previously thought to be 'imperfect fungi' are, in fact, asexual stages in the life-cycle of other documented genera. Earlier botanical texts referred to 'yeasts and yeast-like organisms', which was an indication of the state of flux in classification. True yeasts are unicellular ascomycetous fungi, and, by implication, wild yeasts are also single-celled although some do produce thread-like growths (pseudomycelia) under certain growth conditions. Wild yeasts cause most problems in traditional ale breweries, both during fermentation and when the beer is in the cask. Most of them do not flocculate during fermentation, most are non-reactive with finings, and most produce off-flavours in finished beer.

There are several media available for undertaking total yeast counts, Wallerstein Laboratory Nutrient medium (WLN) being one of the most widely used. WLN supports the growth of brewery organisms generally, but by incorporating a bactericide (tetracycline or chloramphenicol) in the medium, any bacterial growth can be inhibited. To identify *Saccharomyces* spp. from non-*Saccharomyces* spp. generally one can use a defined growth medium that contains lysine as the sole source of nitrogen; members of the genus *Saccharomyces* will not grow. To differentiate ale yeasts in a lager fermentation one can use MYGP agar (which contains malt extract, yeast extract, glucose and peptone) and incubate it at 37 °C. *Sacch. cerevisiae* (and a few wild yeasts) will grow; lager yeast will not. Conversely, by using a defined growth medium with melibiose as the sole source of carbon, one can ascertain the presence of lager yeast in an ale fermentation. Top-fermenting yeast cannot use melibiose as a carbon source.

MYGP medium with Cu^{2+} incorporated will isolate *Saccharomyces* and non-*Saccharomyces* wild yeasts, as will WLN medium with cycloheximide (actidione).

As is the case in bacteriology, in all mycological identification tests employing agar media and subsequent production of visible colonies, there is a considerable lapse of time before results can be obtained. The problems that this presents have already been mentioned. There are, however, one or two more rapid ways of detecting yeast contaminants, one being for the assessment of wild strains of *Saccharomyces*, which produce phenolic off-flavours and odours. The production of such phenolics is under the control of a single gene, POF 1, which regulates the decarboxylation of wort and beer hydroxycinnamic acids (such as *p*-coumaric, cinnamic and ferulic acids). Wild *Saccharomyces* strains likely to be encountered in the brewery possess this gene and are designated Pof$^+$. Brewing strains do not have the ability to undergo such decarboxylations and are Pof$^-$. By incubating the suspect wild yeast in sterile hopped wort containing a ferulic acid supplement, a resultant aroma of cloves (4-vinylguaiacol) confirms that the yeast is wild (Pof$^+$). Incubation is at 27 °C for 24 hours.

Non-brewing *Saccharomyces* species are probably the commonest wild yeasts encountered in breweries; *Sacch. pastorianus*, *Sacch. ellipsoideus* and *Sacch. diastaticus* are most notable. If found in fermentations they can all cause haze problems because of their poor reaction with beer finings. *Sacch. ellipsoideus* does not react with finings at all. *Sacch. diastaticus* is a nuisance since it has the ability to ferment wort dextrins and lead to super-attenuated beers.

The following genera (some of which are disputed) have been recorded as brewery isolates:

(1) *Brettanomyces*. This genus is required in lambic beer production because of its ability to produce desired acids. In other beer styles it is troublesome – for the same reason. Acid is produced from glucose and particularly from ethanol. These yeasts are alcohol tolerant, some strains being able to survive in environments of up to 18% ethanol (sherry). They have also been isolated from bottled wine and soft drinks. *Brettanomyces* is the imperfect form of *Dekkera*, which is ascosporogenous and has also been isolated from beer and wine.

(2) *Hansenula*. As well as existing as unicells, these yeasts can produce thread-like growths under certain conditions (called pseudohyphae). The spores produced are hat- or saturn-shaped. Several genera have been isolated from foodstuffs and industrial fermentations, the most well-known being *H. fabianii*, *H. anomala* and *H. subpelliculosa*. [*H. subpelliculosa* is able to tolerate very high sugar concentrations (*i.e.* it is osmophilic).] Nitrate is assimilated and growth is strictly aerobic; thus, they only infect beers with some oxygen presence, producing a surface raft of growth, or pellicle. When dispersed, the pellicle gives rise to general turbidity.

(3) *Pichia*. This genus is similar in some ways to *Hansenula*, only these yeasts cannot assimilate nitrate. Some produce true mycelium, and some pseudomycelium. Some species are associated with fruit flies (*Drosophila* spp.), where they appear to be an essential part of their diet. They are strictly aerobic fungi and so growth in beer is in the form of a surface pellicle. In terms of

beer spoilage, *P. membranaefaciens* is the commonest culprit; it is also known from spoiled wine. *P. farinosa* also infects alcoholic beverages, especially sake.

(4) *Candida.* A large and heterogenous group of asporogenous yeasts, there being over 200 species. True mycelium (hyphae) are produced by many strains. Most are associated with animals, but some do occur in wineries and breweries. *C. ingens*, which has been isolated from sherry, appears to be the most commonly encountered brewery form. *C. mycoderma* has also arisen in fermentations.

(5) *Torulopsis.* Like *Candida*, this genus is asporogenous, but does not produce any mycelium or pseudomycelium. Many are thought to be imperfect stages in the life-cycle of other (many as yet unknown) genera. Most can assimilate glucose and, to a lesser extent, ethanol and lactic acid. *T. stellata* is common on over-ripe grapes and in grape juice and has been recorded as a brewery isolate.

(6) *Debaryomyces.* A weakly-fermenting ascomycete that produces two warty ascospores per ascus. They grow poorly in the absence of oxygen, especially the brewery isolate *D. hansenii*, which is also highly halotolerant, being able to grow in the presence of 20% sodium chloride. *Torulopsis candida* is known to be an imperfect stage in the life-cycle of a *Debaryomyces* sp. and can survive in very high sugar concentrations.

(7) *Kloeckera.* A genus of imperfect fungi that undergoes bipolar budding, which gives their cells an apiculate shape. All have the ability to ferment sugars. *Kloeckera* spp. are non-sporing forms of the ascomycetous genus *Hanseniaspora*, the brewery contaminant *K. apiculata* being part of the life cycle of *H. uvarum*. The natural habitat for the genus seems to be spoiled fruit.

(8) *Kluyveromyces.* A perfect genus whose ascospores are heat-resistant, *K. bulgaricus* being identified from canned fruit. Ascospores are very easily liberated from the ascus, unlike *Debaryomyces* where they are not liberated at all. Some species ferment lactose, and *K. fragilis* and *K. lactis* are known from dairy products.

(9) *Rhodotorula.* In laboratory cultures, conspicuous red colonies are formed on solid media due to the production of carotenoids. They do not appear to have any fermentative ability. Most authorities regard *Rhodotorula* spp. as the haploid state of the genus *Rhodosporidium*. *Rhodotorula glutinis*, an assimilatory nitrate reducer (and hence capable of contributing to ATNC in

beer), is known to be the mating type of at least three different *Rhodosporidium* spp.

For an extended treatise on microbiology in the brewery the reader should consult Priest and Campbell.[19]

SPONTANEOUS FERMENTATIONS

Certain beers, notably the Belgian lambics and gueuzes, are traditionally brewed using spontaneous fermentation rather than yeast being pitched into the hopped-wort.

Over the millennia a unique microflora has developed around the valley of the River Senne, near Brussels. This has been contributed to by the topography of the area and its microclimate, as well as general agricultural activities. Traditional lambic and gueuze beers are indigenous to this area, although it is now possible to set up a 'lambic brewery' elsewhere, as long as climatic conditions are comparable and the requisite aerial microflora has been introduced and stabilised (which takes at least five years to accomplish). The microflora is most significant in the cooling room of the brewery where the 'infection' needed for the initiation of fermentation takes place. It has only been possible to set up a modern-day lambic brewery artificially since the microbiology of the lambic process has been fully explained. Needless to say, the air in a lambic brewery will always contain a more diverse and abundant microflora than that from one producing lager or ale.

From lambic beers *per se*, gueuze beers and fruit beers can be produced after a period of secondary fermentation.

Lambic Fermentation

The mash for a lambic beer is like no other and is aimed at providing as many wort constituents as possible (including starches and dextrins) in order to accommodate the growth of the multitude of microorganisms that will be involved in fermentation. A typical grist consists of 30% unmalted wheat and 70% malt. The striking temperature of the liquor is around 55 °C and the initial mash temperature 45 °C. After a prescribed stand period, some turbid wort is removed and hot (90 °C) liquor is added which brings the mash temperature up to about 52 °C. Another stand (and wort removal) and subsequent addition of more liquor at 90 °C raises the mash to 65 °C and this is repeated again so that the mash temperature reaches 72 °C. After a period, the whole mash is heated to 85 °C and the wort is filtered off before being run into the copper. The last remnants of wort are rinsed from the mash with 95 °C liquor.

Once in the copper, the worts are boiled for five to six hours with aged hops (which have lost much of their bittering potential), a 2% α-acid level being considered to be the upper limit. There are three principal reasons for using aged hops:

(1) The reduced α-acid (and subsequently iso-α-acid) levels result in lower antimicrobial activity which permits survival of the many microbes involved in fermentation.
(2) Lambic beers are necessarily soft and low bitterness levels are required.
(3) Aged hops contain higher levels of polyphenols (at least 350 ppm) than fresh ones, and these confer a dry tannin taste which is also a requirement of the style. Elevated levels of polyphenols also promote the complicated and delicate flavour stability of lambic beers.

After boiling, hopped wort is passed through a seive into cooling trays (coolships) which are open to the atmosphere. Wort pH is adjusted to 5.0 during boiling to allow growth of bacteria necessary for the first stages of fermentation. Total wort production time in a traditional lambic brewery is about eight hours, each stage being carried out with the aim of generating sufficient moisture to maintain the natural microflora of the brewery environment. This is especially important in the area where the coolships are situated.

Lambic fermentations technically commence whenever the temperature of the hopped-wort in the coolship has been reduced sufficiently to permit microbes from the 'aerial infection' to establish themselves. When appropriately cool, worts are transferred from the coolship to wood-lined fermenting vessels or wooden barrels, and it is here that fermentation proper is carried out.

Lambic fermentations consist of four overlapping phases (five, if gueuze and/or lambic fruit beers are being produced):

(1) due to Enterobacteriaceae;
(2) due to *Saccharomyces* spp.; alcoholic fermentation;
(3) due to lactic acid bacteria; the acidification (lambic) stage;
(4) maturation, softening, lowering of diacetyl and dimethyl sulfide levels;
(5) if gueuze or fruit beers are being brewed then there has to be a re-fermentation involving *Brettanomyces* spp. and lactic acid bacteria.

The flora of the initial enterobacterial fermentation normally include such species as *Escherichia coli, Klebsiella aerogenes, Citrobacter freundii, Enterobacter cloacae, Enterobacter aerogenes* and *Hafnia alvei*, all apart from *K. aerogenes* being motile. All ferment glucose and some ferment lactose with the production of acid and gas. Breakdown of sugars is by mixed acid fermentation or butanediol fermentation, both of which proceed *via* the EMP reactions shown in Figure 7.7. Gas evolution (mainly carbon dioxide and hydrogen) is copious enough to give much foaming, and excess foam is extruded from the fermenter by an overflow pipe. This is important as some unwanted flavours are also removed in the main gas flow. The natural infection of wort gives low initial numbers of fermenting bacteria, a condition deemed to be essential for classic lambic beers. Attempts to seed these organisms artificially, a practice carried out by more 'commercial' breweries, does not yield the same finished product and is frowned upon by the craft brewer. The enterobacterial phase of fermentation lasts about one month. There will be a few non-maltose fermenting yeasts present as well, but they play a minor role in the

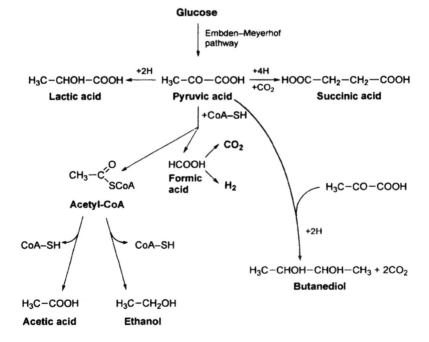

Figure 7.7 *Outline pathways of the mixed acid and butanediol fermentations, as elicited by members of the Enterobacteriaceae*

overall sequence of events. Most commonly encountered at this stage are *Kloeckera apiculata, Saccharomyces dairiensis* and *Sacch. globosus*, all of which are actidione-resistant.

At the termination of this initial phase of fermentation (of about one month in duration), the levels of the desired enterobacterial metabolic end-products will be in the region of:

—lactic acid, 1000 ppm;
—2,3-butanediol, 1000 ppm;
—succinic acid, 250 ppm;
—acetic acid, 200 ppm;
—formic acid, 100 ppm.

Dimethyl sulfide is also formed and can be found at levels of up to 500 ppb.

After four or five weeks, alcoholic fermentation commences when the wort has been suitably modified and *Saccharomyces* species have reached appropriate levels. The most likely species are *Sacch. cerevisiae, Sacch. uvarum, Sacch. inusitatus* and *Sacch. bayanus*. This phase lasts for about three months and, as well as ethanol, a number of esters are liberated, most notably ethyl acetate and ethyl lactate, both of which make important flavour contributions. Fusel alcohols are produced in small amounts and these, together with the organic acid content of the ferment, lead to a wide variety of esterifications. The highest rates of alcohol production are seen between weeks four and fourteen, but alcoholic fermentation occurs slowly throughout the course of the lambic fermentation.

Acetic acid bacteria start to proliferate after about two months of this phase and gradually increase in numbers over the ensuing three months.

After about four months into the fermentation, lambification commences and is due to homofermentative lactobacilli and, more particularly, pediococci. Traditional establishments arrange their brewing regimes so that the warmer summer months coincide with this stage of lambic brewing. This is especially important for pediococci, where a temperature of at least 20 °C is conducive to growth. Important components of the overall lambic flavour profile are introduced by lactics, principally lactic acid, ethyl lactate, diacetyl and acetoin.

During lambification, the pH of the ferment drops below 4.0 and this, together with a gradual reduction in fermentables, heralds the disappearance of *Saccharomyces* species. This is counteracted by the increasing dominance of *Brettanomyces* spp. and the continuation of acidification. Nearly all recognised species of this genus have been recorded, the most abundant being *B. bruxellensis* and *B. lambicus*. These two actidione-resistant yeasts play a vital role in the regulation of flavour-active compounds, and the best lambic beers result from an acidification with an early dominance by *B. bruxellensis* and a later, more prolonged, presence of *B. lambicus*.

Iso-amyl acetate, which is considered undesirable, is removed by *Brettanomyces* by means of a cell-bound esterase, but, overall, the ester composition of the brew is increased; ethyl acetate and ethyl lactate are vigorously synthesised. A number of new metabolites are introduced by *Brettanomyces* activity, the most important being iso-butyric acid, caproic acid, caprylic acid, capric acid, ethyl caprylate and ethyl caprate.

Total time taken for lambification/acidification is around six months, thus taking the whole fermentation period up to around ten months. There is relatively little fermentative activity after this time, but numerous maturation reactions are undertaken during subsequent storage. In terms of flavour, reduction of diacetyl and dimethyl sulfide are important; levels being reduced to 80 and 100 ppb, respectively. The numbers of viable organisms are reduced during maturation, especially those of lactic acid bacteria and *Brettanomyces* spp. The normal maturation time for a lambic is 24 months, although 'young' lambic is produced which is only matured for one year and which still has a discernible carbohydrate content. Finished lambic beers have a pH of close to 3.25 and, by definition, should not have a pH of over 3.8. The changes in population of some of the major groups of micro-organism involved in a spontaneous lambic fermentation are shown in Figure 7.8.

Gueuzes

From a straight lambic beer, gueuzes and lambic fruit beers are brewed by means of a natural secondary fermentation, usually in-bottle. Gueuzes are formed by blending a young lambic beer (less than one year old) with an aged one (at least three years old). The young beer contains an active yeast population, whilst the old beer contains dextrinases that have originated from some of the microbes partaking in the overall fermentation. When mixed together, dextrinases break down residual macromolecular carbohydrates to fermentable sugars, which are then slowly attacked by the yeast flora. The ratio of blending is two parts young lambic to one part old. The normal residence time in-bottle is about 15 months, after which time a

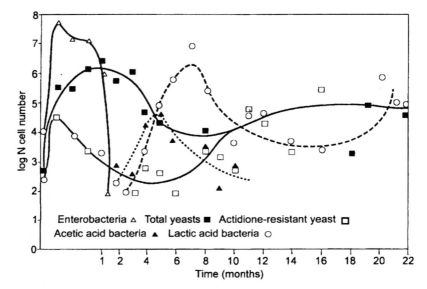

Figure 7.8   *Changes in microbiological populations during spontaneous lambic*
fermentation
(Courtesy of the *Brewers' Guardian*[20])

sparkling, astringent gueuze beer is produced. Some authorities describe gueuzes as being champagne-like.

At the point of bottling, the majority of the microflora is in the
form of yeasts and lactic acid bacteria, both being present at levels of
approximately 10^2 cells per ml. After one month in-bottle yeast counts
will have risen to 10^5 per ml, and lactics to 10^6 per ml as available
carbohydrates are metabolised. The dominant yeast genus is *Brettanomyces*, although actidione-sensitive genera such as *Torulopsis, Hansenula, Candida, Cryptococcus* and *Pichia* are detectable for the first few
months in-bottle. After 14 months of re-fermentation, only lactic acid
bacteria, particularly pediococci, are viable to any extent.

Fruit Beers

The best fruit beers are based on lambics, the original ones being
cherry-krieks which were brewed every July, in Belgium, to coincide
with the cherry season. Traditionally, crushed cherries would be
introduced into wooden casks containing a young (12 month) lambic
beer. Sugars and micro-organisms from the surface of the cherries
would stimulate further fermentation. Nowadays, lambic fruit beer

production need not be quite so seasonal because the fruit can be kept frozen and juice extracted after thawing before addition to young lambic. Fruit beers constructed in this way have more colour and a slightly different flavour than those made with freshly-picked fruit, but they are accepted by traditional lambic brewers.

REFERENCES

1 Institute of Brewing, *Methods of Analysis, Vol. 1 – Analysis*, January 1997.
2 Institute of Brewing, *Methods of Analysis, Vol. 2 – Microbiological*, January 1997.
3 C.H. Collins and P.M. Lyne, *Microbiological Methods*, Butterworths, London, 3rd edn., 1970.
4 J.R.M. Hammond, *The Brewer*, 1996, **82**, 332.
5 R.J. Stewart and T.M. Dowhanick, *J. Am. Soc. Brew. Chem.*, 1996, **54**, 78.
6 L. Laidlaw, T.A. Tompkins, L. Savard and T.M. Dowhanick, *J. Am. Soc. Brew. Chem.*, 1996, **54**, 97.
7 G. Vogeser and E. Geiger, *Brauwelt*, 1998, **138**, 1060.
8 J.L. Shimwell and M. Grimes, *J. Inst. Brew.*, 1936, **42**, 348.
9 F.B. Strandskov and J.B. Bockelmann, *Wall. Lab. Commun.*, 1955, **18**, 275.
10 J.L. Shimwell, *Brewers' J.*, 1963, **99**, 759.
11 F.G. Priest, H.J. Somerville, J.A. Cole and J.S. Hough, *J. Gen. Micro.*, 1973, **75**, 295.
12 D.J. Brenner, in *The Prokaryotes; a Handbook on Habits, Isolation and Identification of Bacteria*, eds. M.P. Starr, H. Stolp, H.G. Truper, A. Balows and H.G. Schlegel, Springer-Verlag, New York, 1981. pp. 1105–1127.
13 J. Swings and J. DeLey, in *Bergey's Manual of Systematic Bacteriology, Vol. 1*, ed. N.R. Kreig, Williams and Wilkins, Baltimore, 1984, p. 576.
14 S.Y. Lee, M.S. Mabee and N.O. Jangaard, *Int. J. Syst. Bacteriol.*, 1978, **28**, 582.
15 N. Weiss, H.Seidel and W. Back, *Brauwissenschaft*, 1979, **32**, 189.
16 U. Engelmann and N. Weiss, *Syst. Appl. Microbiol.*, 1985, **6**, 287.
17 F. Eschenbecher, *Brauwissenschaft*, 1968, **21**, 424.
18 W. Back, I. Bohak, M. Ehrmann, W. Ludwig and K.-H. Schleifer, *Syst. Appl.Microbiol.*, 1996, **19**, 322.
19 F.G. Priest and I. Campbell, *Brewing Microbiology*, Chapman and Hall, London, 2nd edn., 1996.
20 R.A. Mussche, *Brewers' Guardian*, 1999, **128**, 19.

Subject Index